Hydrate Engineering

E. Dendy Sloan Jr.
Director, Center for Research on Hydrates
Colorado School of Mines

Edited by

J. Ben Bloys
Principal Research Engineer
Arco E&P Technology Inc.

Henry L. Doherty Memorial Fund of AIME
Society of Petroleum Engineers Inc.

Richardson, Texas

Dedication

This monograph is dedicated to my wife Marjorie, whose magnanimous patience and good humor never failed to cast a cheerful glow.

Printed in the United States of America.

ISBN 978-1-55563-088-1
ISBN 978-1-61399-315-6 (Digital)

Society of Petroleum Engineers
222 Palisades Creek Drive
Richardson, TX 75080-2040 USA

http://www.spe.org/store
books@spe.org
1.972.952.9393

SPE Monograph Series

The Monograph Series of the Society of Petroleum Engineers was established in 1965 by action of the SPE Board of Directors. The Series is intended to provide authoritative, up–to–date treatment of the fundamental principles and state of the art in selected fields of technology. The Series is directed by the Society's Monograph Committee. Below is a listing of those who have been most closely involved with the preparation of this monograph.

Monograph Committee (2000)

Disclaimer

The descriptions, methods, and cases discussed in this monograph are presented solely for educational purposes and are not intended to constitute design or operating guidelines or specifications. While every effort has been made to present current and accurate information, the author (and sponsoring and contributing organizations) assume no liability whatsoever for any loss or damage resulting from the use of this material or for any infringement of patents or violation of any federal, state, or municipal regulations. This book is intended to supplement, but not replace, engineering judgment. Use of the information herein is solely at the risk of the reader.

Acknowledgments

Arco E&P Technology Co. is acknowledged gratefully for its funding of *Hydrate Engineering*. The foundation of the present work was made possible by Arco's kind permission to make the previously proprietary handbook public.

This work is an extension of the *Offshore Hydrate Engineering Handbook* (Arco E&P Technology Co., January 1998), an idea conceived by Ben Bloys of Arco E&P Technology Co. This work is a paean to Bloys' foresight regarding the state of hydrate engineering knowledge, coupled with intelligence and a magnanimous perspective. Bloys' continual input, encouragement, and editorship of this work were essential to its publication.

Two others have been fundamental to the project. Jim Chitwood of Texaco ensured Deepstar hydrate-related reports (Phases I, II, and IIA) were made available to this project. The power of a multicompany consortium, demonstrated by DeepStar's approach to the difficult offshore flow-assurance problem, has provided an invaluable supplement to the manuscript. All the DeepStar figures and report summaries were approved for inclusion in this monograph.

John Cayias, formerly of Oryx Energy, contributed by providing visits to offshore platforms and by providing funds for Matt Seefeldt, the student worker who aided in production of the figures. Cayias' questions have been very useful in rethinking and restating the concepts summarized in this book.

The following people have contributed generously.

BP-Amoco's George Shoup and J.J. Xiao provided hydrate-plug transient-flow-simulation results and reviewed the preliminary draft.

At Arco, in addition to Bloys' multiple contributions, particularly in the areas of drilling and well testing, Phil Lynch kindly provided the most detailed North Sea case study. Greg Walz contributed to the drilling-fluid portion of the book.

Johnny Johnson of Black, Veatch, and Pritchard provided experience and diagrams of onshore process points where hydrates commonly form.

British Petroleum contributed heavily through Carl Argo, Chris Osborne, and particularly Tony Edwards, who related North Sea commercial offshore operating experiences with new inhibitors.

Chevron's Pat Shuler generously contributed his spreadsheet program HYDCALC to determine inhibition amounts and provided access to offshore engineers. Karl Gerdes reviewed the guidelines for safety, design, and operation.

Conoco's Stan Swearingen and Mobil's Barry Hoffman were helpful in reviewing guidelines and manuscript drafts.

Bill Parrish of Phillips provided a hydrate perspective gained over a quarter century of research and plant optimization. His collaboration provided an essential bridge between the theoretical and industrial perspectives. Parrish served as the second reviewer for this manuscript.

The hydrate team composed of T. Austvik, L.H. Gjertsen, O. Urdahl, and A. Lund at Statoil's Research Center in Trondheim provided 2 full days of interviews regarding hydrate prevention and remediation in the Norwegian sector of the North Sea.

At Texaco, in addition to Chitwood's connection with Deepstar, Phil Notz has been a hydrate colleague for more than a decade. He provided information on inhibitor economics, feedback on guidelines, and reviewed the draft of the manuscript. Jack Todd, the principal author of the *Texaco Reliability Engineering Manual* for operating personnel, was extremely helpful in providing the book and in arranging interviews with Texaco engineers.

John Barker of Vastar related his pioneering drilling experiences with hydrates in drilling and testing wells. His former Exxon colleague, Janardhan Davalath of FMC was helpful in providing completion and well testing diagrams.

The efforts of these technical personnel have been invaluable in moving the handbook to industrial utility. However, this book utimately is the responsibility of (and is limited by) a personal perspective, intended to assimilate and synthesize the previously mentioned contributions and those in the laboratory and literature.

Preface

Natural gas hydrates are crystals formed by water with natural gases and associated liquids in a ratio of 85 mol% water to 15% hydrocarbons. The hydrocarbons are encaged in ice-like solids that do not flow but rapidly grow and agglomerate to sizes that can block flowlines. Hydrates can form anywhere and anytime that hydrocarbons and water are present at the right temperature and pressure, such as in wells, flowlines, or valves and meter discharges. Appendix A gives hydrate crystal details at the molecular level, along with the similarities to and differences from ice.

The low temperatures and high pressures of the winter and deepwater environments promote hydrate formation as a function of gas and water composition. In a pipeline, hydrate masses usually form at a hydrocarbon/water interface and accumulate as flow pushes them downstream. The resulting porous hydrate plugs have the unusual ability to transmit some degree of gas pressure while they hinder flow. Frequently, both gas and liquid can be transmitted through the plug; however, lower viscosity and surface tension favor the flow of gas. Pipeline depressurization, which sometimes prevents normal production for weeks, is the principal tool for hydrate-plug removal.

This monograph provides operating and design engineers with practical answers to the following four questions.

- What are the safety problems associated with hydrates? (Chap. 1)
- What are the best methods to prevent hydrates? (Chap. 2)
- How are hydrate plugs best removed? (Chap. 3)
- What are the economics of prevention and remediation? (Chap. 4)

Field case studies, pictures, diagrams, and example calculations are the basis for this book. Appendix A provides background material on hydrate structures and formation mechanism. A computer program disk and User's Guide (Appendix B) are provided to enable prediction of hydrate conditions. Appendix C is a compilation of case studies not included in the text, while Appendix D is a compilation of all the rules of thumb provided in the text, with references, to the appropriate sections.

For additional references, Makogon[1,2] presents a Russian hydrate perspective and Sloan[3] gives an in-depth hydrate treatment.

References

1. Makogon, Y.F.: *Hydrates of Natural Gas,* PennWell Publishing Co., Tulsa, Oklahoma (1981).
2. Makogon, Y.F.: *Hydrates of Hydrocarbons,* PennWell Publishing Co., Tulsa, Oklahoma (1997).
3. Sloan, E.D. Jr.: *Clathrate Hydrates of Natural Gases,* Marcel Dekker Inc., New York City (1998).

Table of Contents

Chapter 1
Safety First: A Gallon of Prevention Is Worth a Mile of Cure

1.1 Introduction

This chapter presents case histories of some of the dangers associated with hydrate plugs. Many instances of line rupture, sometimes accompanied by loss of life, are attributed to the formation of hydrate plugs. Three characteristics of hydrates contribute to safety problems.

1. Hydrate density is similar to ice, and upstream pressure can propel a dislodged hydrate plug at high velocity. In 1997 DeepStar Wyoming field tests, plug lengths ranged from 25 to 200 ft with velocities between 60 and 270 ft/sec. As **Fig. 1.1** shows, such velocities and masses provide sufficient momentum to cause two types of failure at a pipeline restriction (orifice), obstruction (flange or valve), or a sharp change in direction (bend, elbow, or tee). First, hydrate impact can fracture pipe, and, second, extreme gas compression can cause pipe rupture downstream of the hydrate path.

2. Hydrates can form either single or multiple plugs, with no method to predict which will occur. High differential pressures can be trapped between plugs, even when the discharge ends of plugs are depressurized.

3. Hydrates contain as much as 180 volumes of gas at standard temperature and pressure (STP) per volume of hydrate. When hydrate plugs are dissociated by heating, any confinement causes rapid gas-pressure increases. However, heating is not an option within buried pipe because of the difficulty of locating the plug and the economics of heating an inaccessible pipeline.

Field engineers discuss the "hail-on-a-tin-roof" sounds that small hydrate particles make when striking a pipe wall. Such small, mobile particles can accumulate into large masses that occupy a considerable volume, often filling tens or hundreds of feet of the pipeline length. Attempts to blow the plug out of the line by increasing upstream pressure (see Appendix D, Rule-of-Thumb 19) result in additional hydrate formation and perhaps pipeline rupture.

When a plug is depressurized with a high differential pressure, the dislodged plug can be a very dangerous projectile, as Secs. 1.2.1 through 1.2.3 indicate.*

1.2 Single-Sided Plug Depressurization Hazards

1.2.1 Case Study 1.1—1991 Chevron Incident. A foreman and an operator were attempting to clear a hydrate plug in a sour-gas flowline. They had bled down the pressure in the distant end from the wellhead and were standing near the line when the line failed, probably from the impact of a moving hydrate mass. A large piece of pipe struck the foreman. Although an air ambulance was deployed, the foreman was declared dead on arrival at the hospital. No pre-existing pipe defects were found. See Case Study 3.8 for a similar incident involving the loss of life.

1.2.2 Case Study 1.2—1991 Gulf Oil Incident. On 10 January 1991, the Rimbey gas plant was in the startup mode. A hydrate or ice plug formed in the overhead line from the amine contactor. The line had been depressured to the flare system, downstream of the plug. The ambient temperature, which had been −30°F, rose rapidly owing to warming winds around midnight. At 2 a.m. the overhead line came apart, killing the chief operator. In addition, the plant suffered approximately U.S. $6 million in damage.

Pre-existing cracks in the pipeline contributed to this failure. While these cracks did not impair the piping's pressure-containing ability under steady-state conditions, they reduced the piping strength under the transient (impact) conditions when the plug broke free.

1.2.3 Case Study 1.3—1991 Mobil Incident. At 11:30 a.m. on 2 January 1991, two operators attempted to remove a blockage in a sour-gas flowline, which had been plugged for about 3 days. The downstream side of the plug had been depressured completely. The upstream portion of the line, originally at 1,100 psig, was depressured completely to a truck within a 5-minute period. At 12:15 p.m., the flowline failed and gas began flowing from somewhere around the casing. An employee of a well-control/firefighting company isolated the leak at 3:18 p.m.

The eruption of a hydrate plug at a Schedule 40, 3-in., screwed-pipe nipple caused the failure. Because both ends of the hydrate plug were depressured, two end plugs may have existed, with intermediate plugs (**Fig. 1.2a**).

1.2.4 Common Safety-Hazard Conditions. In Case Studies 1.1 through 1.3, common equipment circumstances existed. The systems exhibited the following characteristics.

1. They were out of service immediately before the incident.
2. They did not have hydrate or freeze protection.
3. They were pressurized while out of service.
4. They were being restarted.

*Kent, R.P. and Coolen, M.E.: "Hydrates in Natural Gas Lines," internal report, Mobil (1992).

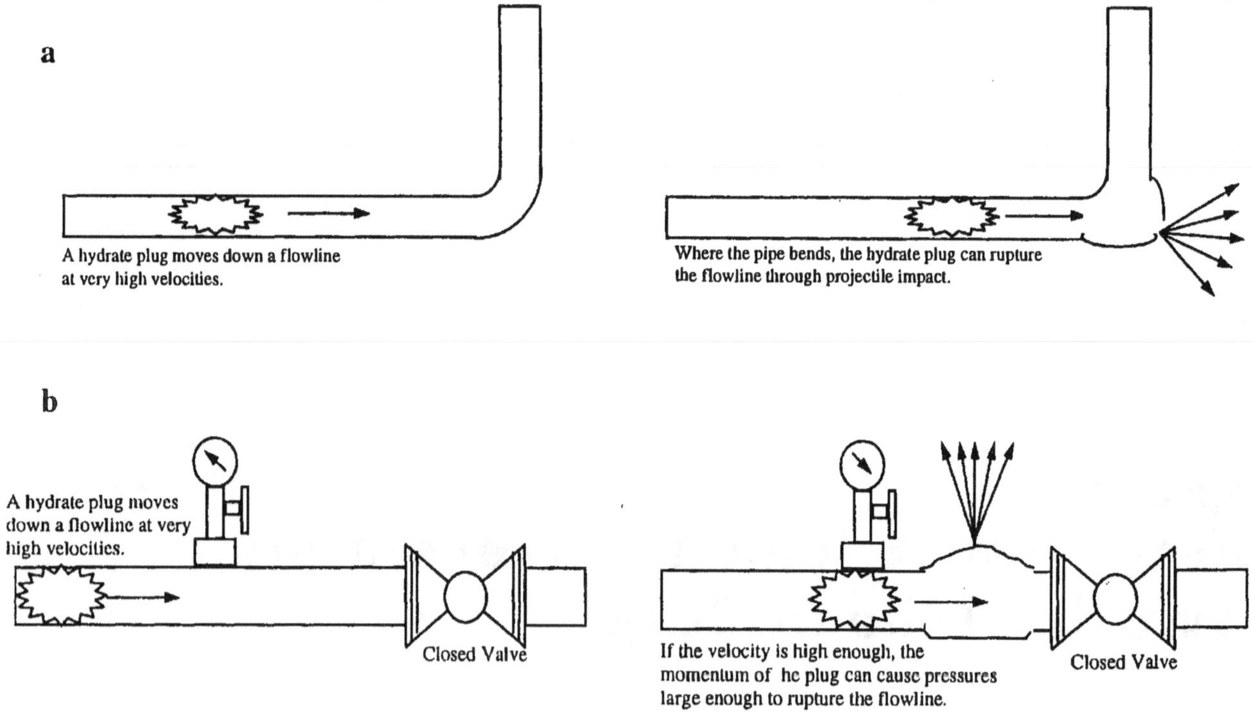

Fig. 1.1—Safety hazards for moving hydrate plugs [from an internal report, *Hydrate Handling Guidelines: Safety and Loss Control Manual*, Chevron Canada Resources (23 July 1992)].

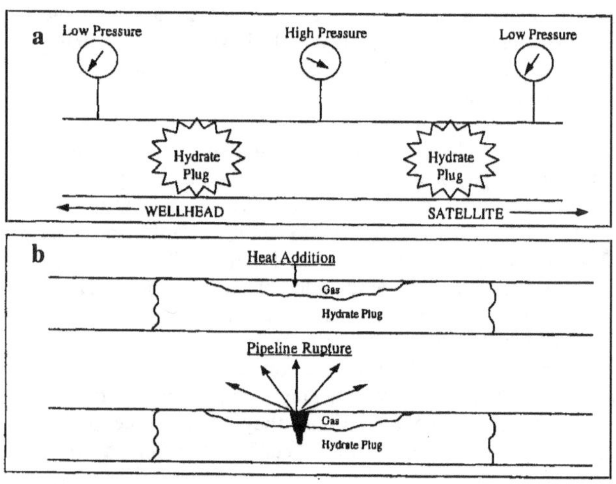

Fig. 1.2—Safety hazards of high pressures trapped by hydrates [from an internal report, *Hydrate-Handling Guidelines: Safety and Loss Control Manual*, Chevron Canada Resources (23 July 1992)].

5. They had high differential pressures across plugs for short periods.

The Chevron Canada Resources' *Hydrate-Handling Guidelines** suggest that the danger of line failure owing to hydrate plug(s) is more prevalent when (1) long lengths of pressurized gas are trapped upstream, (2) low downstream pressures provide less cushion between a plug and restriction, and (3) restrictions/bends exist downstream of the plug.

1.3 Hazards of Applying Heat To Remove Hydrate Plugs

1.3.1 Case Study 1.4—Statoil North Sea Incident. In the mid-1980's, a hydrate plug occurred topside on a platform in a Statoil oil field in the Norwegian sector of the North Sea. The line section was valved off, and heat was applied to remove the plug. After some heating time, the work crew went to lunch, intending to complete the task on their return. When they returned, the crew found that the section of line had exploded during their absence.

Heat had been applied to the midpoint of hydrate plug, and the plug-end portions contained very high pressures until the line ruptured. **Fig. 1.2b** shows a schematic of such a situation. Chap. 2 shows that pressure increases exponentially with increasing temperature when hydrates are dissociated.

1.4 Hazards of Expelling Plugs With High Upstream Pressure

1.4.1 Case Study 1.5—Elf Aquitaine North Sea Incident. In the 1970's, a plug occurred on a floating platform riser in the North Sea. Blocking valves were closed and the pipeline was disconnected downstream of the plug. The discharge end of the pipeline was aimed overboard with the intent of using high upstream pressure to extrude the plug from the line. When the plug was expelled into the ocean, the force was so great that observers say the platform rose 8 in. in the ocean.

1.5 General Hydrate Plug Safety Considerations

The Canadian Assn. of Petroleum Producers' (CAPP) hydrate guidelines[1] suggest three safety concerns in dealing with hydrate blockages.

1. Always assume multiple hydrate plugs; there may be pressure between the plugs.
2. Attempting to move ice (hydrate) plugs can rupture pipes and vessels.
3. While heating a plug normally is not an option for a buried pipeline, any heating should be done from the end of a plug, rather than the middle.

The third recommendation could be expanded in consideration of a buried line: (1) heating a submerged plug is not recommended because of the inability to determine the end of the plug as well as provide for gas expansion on plug heating and (2) depressuring a plug gradually from both ends is recommended.

**Hydrate-Handling Guidelines: Safety and Loss Control Manual*, internal report, Chevron Canada Resources (23 July 1992).

1.6 Conclusions

These case studies warn that hydrates can be hazardous to personnel and to equipment. Yet hydrate plugs can be dissociated safely with the procedures in Chap. 3, Remediation.

The preferred procedure, from both safety and economic considerations, is to prevent the formation of hydrate plugs through design and operating practices. While the continuous use of many gallons of inhibitors may be costly, such expenses are overshadowed easily when plugs form and production stops. As the case studies show, it is not uncommon for several hundred yards of hydrate plugs to form, preventing production for a matter of weeks or months during remediation.

Reference

1. King, R. *et al.*: *CAPP Guidelines for the Prevention and Safe Handling of Hydrates,* Canadian Assn. of Petroleum Producers, Calgary (January 1994).

SI Metric Conversion Factors

ft	× 3.048*	E − 01	= m
°F	(°F−32)/1.8		= °C
in.	× 2.54*	E + 00	= cm
psi	× 6.894 757	E + 00	= kPa

*Conversion factor is exact.

Chapter 2
Prevention by Design: How To Ensure That Hydrates Won't Form

2.1 Introduction

The purpose of this chapter is (1) to indicate common sites of hydrate formation, (2) to indicate design methods to provide hydrate protection, and (3) to provide designs that make remediation easier if a hydrate plug occurs.

Hydrates form in hydrocarbon processes in two fundamental ways: slow cooling of a fluid as in a pipeline (see Example 2.2) or rapid cooling caused by depressurization across valves or through a turboexpander (see Example 2.3). Sec. 2.2 describes the conditions under which hydrates form, and Sec. 2.3 gives six prevention methods. Sec. 2.4 provides typical examples of hydrate formation in a well, a flowline, a platform, and an onshore process. Secs. 2.5 through 2.7 describe the design for hydrate thermodynamic inhibition with slow cooling of a pipeline. Sec. 2.8 provides design practices for hydrate prevention with rapid cooling across a restriction like a valve. Sec. 2.9 gives procedures for preventing hydrates through inhibition and heat management. Sec. 2.10 states drilling practice to prevent hydrate formation. Finally, Sec. 2.11 summarizes general design guidelines for hydrate prevention in hydrocarbon processes.

2.2 Hydrate-Formation Conditions

Three conditions promote hydrate formation in processes.

1. Free water and natural gas components must be present. Gas molecules ranging in size from methane to butane are typical hydrate components, including CO_2, N_2, and H_2S. The water in hydrates can come from free water produced from the reservoir or from water condensed by cooling the hydrocarbon fluid. The pipeline residence time is usually insufficient for a blockage of hydrates to form from water remaining vaporized in the gas or water dissolved in a hydrocarbon liquid.

2. Low temperatures normally are needed for hydrate formation; however, even though hydrates are 85 mol% water, the system temperature does not need to be below 32°F for these ice-like solids to occur. Onshore pipelines usually experience hydrate conditions during winter months. Offshore, however, below approximately 3,000 ft of water depth, the ocean-bottom (mudline) temperature is remarkably uniform at 38 to 40°F and pipelined gas readily cools to this temperature within a few miles of the wellhead. Hydrates can form easily at temperatures higher than 70°F at high pressures as well.

3. High pressures commonly promote hydrate formation. At 38°F, common natural gases form hydrates at pressures as low as 100 psig; at 1,500 psig, common gases form hydrates at 66°F. Because pipelines typically operate at higher pressures, hydrate prevention is a primary consideration for flow assurance.

2.3 Thermodynamic Prevention Methods

The three conditions necessary for hydrate formation lead to four classic thermodynamic prevention methods.

1. Water removal provides the best protection. Separation removes free water. Water condensation from the gas phase is prevented by drying the gas, either with triethylene glycol to obtain water content of less than 7 lbm/MMscf or with molecular sieves to obtain lower water content. Water removal may not be feasible between the wellhead and the gathering point (or platform), so other prevention schemes, such as inhibitor injection (see Method 4), must be used.

2. Maintaining high temperatures keeps the system in the hydrate-free region (see Sec. 2.9.4). High reservoir fluid temperature may be retained through insulation and pipe bundling, or heat may be added through hot fluids or electrical heating, although this latter option is not economical in many cases.

3. The system pressure may be decreased below the hydrate-formation pressure. To do so, one can design system-pressure drops at high-temperature points (e.g., bottomhole chokes). The resulting lower density, however, decreases pipeline-transportation efficiency.

4. Hydrate prevention is achieved most frequently by injecting an inhibitor, such as methanol (MeOH) or monoethylene glycol (MEG), which act as antifreezes, to decrease the hydrate-formation temperature below the operating temperature.

Two new means of hydrate inhibition have been added to the thermodynamic inhibitor list and are being brought into common practice.

5. Kinetic inhibitors are low-molecular-weight polymers and small molecules dissolved in a carrier solvent and injected into the water phase in pipelines. These inhibitors bond to the hydrate surface and prevent significant crystal growth for a period longer than the free-water residence time in a pipeline. Condensate may or may not be present for this prevention method to be effective. Water and small hydrate crystals are removed at a platform or onshore.

6. Antiagglomerants are dispersants that cause the water phase to be suspended as small droplets in the oil or condensate. When the suspended water droplets convert to hydrates, pipeline flows are maintained without blockage. Because this prevention method relies on

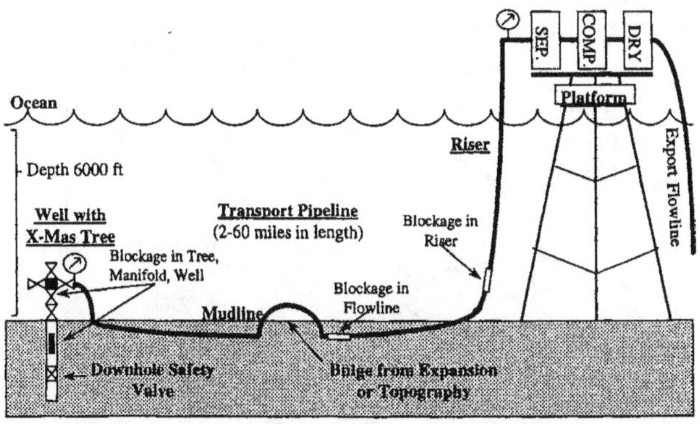

Fig. 2.1—Offshore well, transport pipeline, and platform.

emulsified water/hydrates, a condensed hydrocarbon is required. The emulsion is broken, and water is removed onshore or at a platform.

These methods are used individually or jointly for prevention. This chapter provides ways to use these six methods to prevent hydrates in the system design.

2.4 Where Do Hydrates Form?

Fig. 2.1 shows a simplified offshore process between the subsea-well inlet and the platform-export discharge where virtually all hydrate problems occur. The figure shows hydrate blockages in susceptible portions of the system: (a) the well, (b) the pipeline, or (c) the platform; Examples 2.1, 2.2, and 2.3, respectively, provide brief descriptions of each. High reservoir temperatures prevent hydrate formation prior to the well when the system is flowing. After the platform, export lines have dry gas and oil/condensate with insufficient water to form hydrates.

In Fig. 2.1, two unusual aspects of the system should be noted: (1) the water depth is shown as 6,000 ft, but it may range to 10,000 ft, and (2) the distance between the well and the platform may range to more than 60 miles. Such depths and distances provide cooling for the pipeline fluids to low temperatures and high pressures, which are well within the hydrate-stability region. At a typical ocean temperature of 39°F, 400 ft of water depth provides pressures required for hydrate formation.

The system temperature and pressure at the point of hydrate formation must be within the hydrate-stability region, as determined by the methods in Secs. 2.5 through 2.7. The system temperature and pressure enters into the hydrate-formation region either through a normal cooling process (Example 2.2) or through a Joule-Thomson process (Sec. 2.8).

Fig. 2.2 shows a typical plot of the water temperature in the Gulf of Mexico as a function of water depth. The plot shows that a high temperature of 70°F (or more) occurs for the first 250 ft of depth. However, when the depth exceeds 3,000 ft, the bottomwater temperature is very uniform at approximately 40°F, no matter how high the temperature is at the air/water surface. This remarkably uniform water temperature at depths greater than 3,000 ft occurs in almost all the Earth's oceans (caused by water-density inversion), except a few that have cold subsea currents.

The ocean acts as a heat sink for any gas or oil produced so that, without insulation or other heat-control methods, any flowline fluid cools to within a few degrees of 40°F within a few miles of the wellhead. The rate of cooling with length is a function of the initial reservoir temperature, the flow rate, the pipeline diameter, and other fluid-flow and heat-transfer factors. As Sec. 2.5 explains, however, the ocean-bottom temperature of 40°F is low enough to cause hydrates to form at any typical pipeline pressure.

Example 2.1: Hydrate Formation in a Well. Fig. 2.3 shows a typical subsea well in which fluids are produced through the wing valve and choke to the pipeline. A pressure indication just beyond the choke is essential for determination of hydrate formation in the connecting flowline. Approximately 300 to 500 ft below the mudline is the downhole safety valve (DSV), which is used as the initial emergency barrier between the reservoir and the production system. At the top of the well are swab valves, which provide an entryway into the well for lubricating hydrate-dissociation tools (e.g., inhibitor injection, heaters, coiled tubing) to reach any hydrate blockage.

Hydrate formation in wells is an abnormal occurrence, arising during drilling, testing, or shut-in/startup of the well. While this example provides a brief overview, Sec. 2.10 details methods for prevention of hydrate formation in well drilling and testing. Addressing well blockages may be done with techniques discussed in Chap. 3, Remediation. Case Studies 3.3 and 3.8 provide two additional hydrate remediation experiences in a well.

Davalath and Barker[2] provide a comprehensive set of conditions for dealing with hydrates in deepwater drilling and testing, including two case studies of problems (summarized in Appendix C Case Studies C.23 and C.24) and four case studies of successful hydrate management. Typically, MeOH-injection capability is provided in the well at two places: at the subsea tree and downhole several thousand feet below the seafloor. The injection location and amount of MeOH injected are determined with the procedure in Sec. 2.9.1 on MeOH injection.

In offshore well drilling, a water-based drilling fluid frequently is used that can form hydrates and plug blowout preventers (BOP's), kill lines, and other such equipment, when a gas bubble (or "kick") comes into the drilling apparatus. This represents a potentially dangerous situation for well control. Hydrate formation during drilling is an area of active research with several joint-industry projects.

Barker[3] indicates the following rules of thumb used by Exxon in considering hydrate formation with drilling fluids.

• Drilling hydrate problems occur frequently, but have been recognized only in recent years.

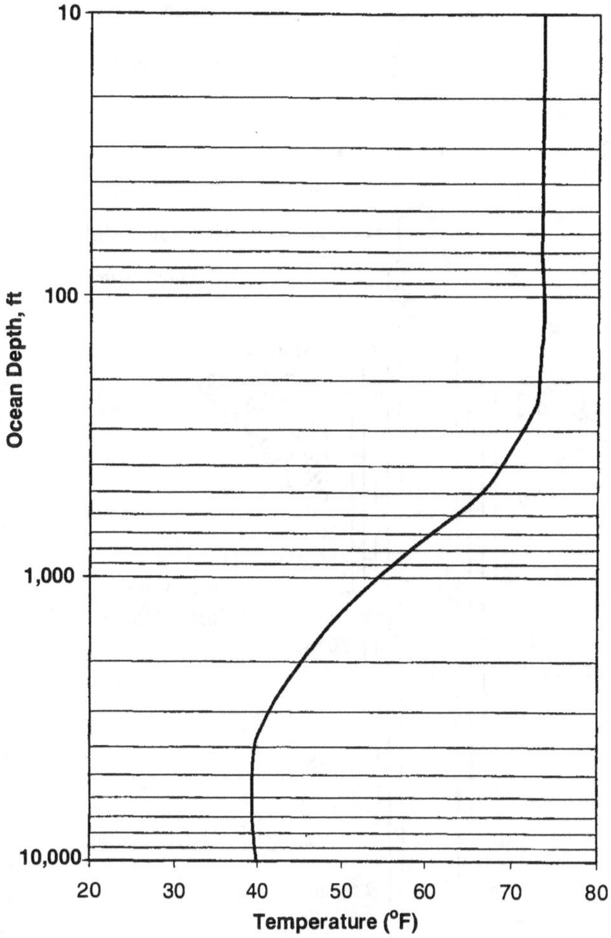

Fig. 2.2—Water temperature vs. depth, Gulf of Mexico.[1]

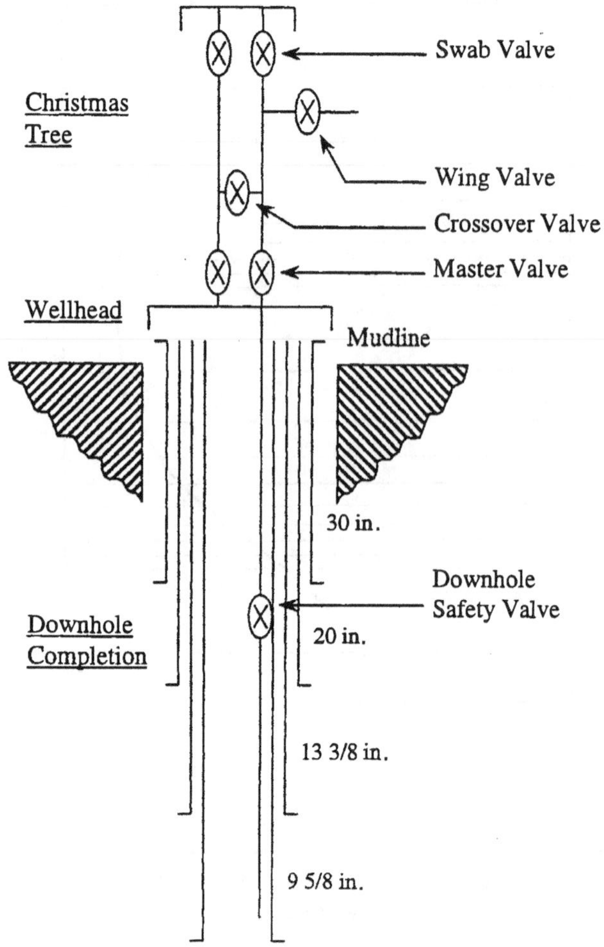

Fig. 2.3—Typical subsea well.

• When hydrates form solids, they remove water from the mud, leaving a solid barite plug.

• A well should not be designed that operates only outside the hydrate region if flow conditions are maintained. If the well is in the hydrate-formation region at static conditions, flow stops at some period and well operation is jeopardized.

• Several hours may be required for hydrate formation and blockage to occur.

• As of October 1988, the company used salt at the saturation limit range of 150 to 170 g/L to prevent hydrate formation.

• **Table 2.1** gives a summary of general guidelines concerning hydrate formation at various water depths.

By 1988 Shell had drilled 16 wells in the Gulf of Mexico at water depths between 2,000 and 7,500 ft using muds with 20 wt% sodium chloride (NaCl) and partially hydrolyzed polyacrylamide (PHPA). In each well, the company experienced an average of more than one gas kick per well, which signaled the possibility of hydrate formation. Only one instance in 2,900 ft of water involved the possibility of hydrate formation, when Shell experienced difficulty disconnecting the drill stack.

TABLE 2.1—GUIDELINES FOR DEEPWATER HYDRATE FORMATION IN DRILLING MUDS IN WATER-BASED MUDS[3]

Water Depth (ft)	Risk of Hydrate-Formation Problems
<1,000	A hydrate problem probably will not occur
1,500	Without inhibition, a hydrate problem may occur
2,000	Without inhibition, a hydrate problem will occur
3,000	Insufficient experience; salt alone will not suffice

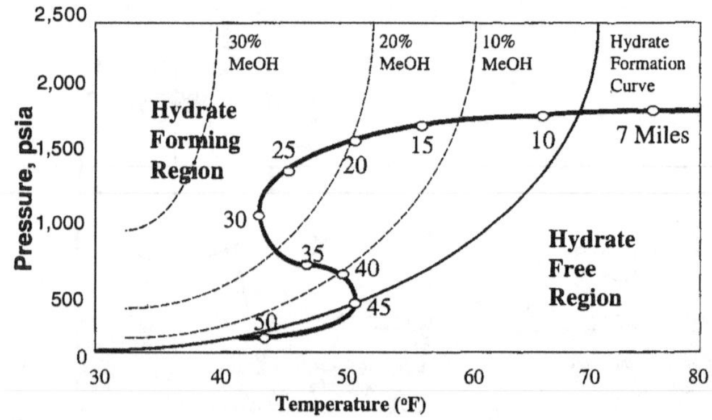

Fig. 2.4—Offshore pipeline plotted on hydrate-formation curves.[5]

Barker and Gomez[4] documented two occurrences (see Case Studies C.21 and C.22 of Appendix C) of hydrate formation in relatively shallow waters off California and the Gulf of Mexico, where losses in drill times were 70 and 50 days, respectively. Recently the number of hydrate problems have increased dramatically with deeper-water drilling. In several cases where safety was an issue (such as plugged BOP's and stack connectors), wells were abandoned. Much remains to be done. Sec. 2.10 specifies some resolved details.

Downstream of the well and choke, the fluid flows through a pipeline of considerable length before reaching either the platform or the onshore process. Example 2.2 represents flow conditions in the pipeline.

Example 2.2: Hydrate Formation in a Pipeline. Notz[5] provides a pipeline case of hydrate formation in **Fig. 2.4** for a Gulf of Mexico gas. On the right of the diagram, hydrates do not form and the system exists in the fluid (hydrocarbon and water) region. However, hydrates form in the shaded region on the left of the diagram, and hydrate-prevention measures should be taken.

Pipeline pressure and temperature conditions were predicted with a pipe-prediction program, such as OLGA® or PIPEPHASE®, and those conditions are shown superimposed on the hydrate conditions in Fig. 2.4. At short pipeline distances (e.g., 7 miles), the flowing stream retains a high temperature from the hot reservoir fluids at the pipeline entrance. The ocean cools the flowing stream, and, at approximately 9 miles, a unit mass of flowing gas and associated water enters the hydrate region (shaded region to the left of the line marked 0% MeOH) and remains in the uninhibited hydrate area until Mile 45. Such a distance may represent several days of residence time for the water phase so that hydrates would undoubtedly form if inhibition steps were not taken.

In Fig. 2.4, the temperature of the pipeline system is within a few degrees of the ocean-floor temperature by Mile 25, so that approximately 23 wt% MeOH is required in the free-water phase to prevent hydrate formation and subsequent pipeline blockage. MeOH-injection facilities are not available at the needed points (miles 9 to 45) along the pipeline. Instead, MeOH is injected into the pipeline at the subsea wellhead. In the case of the pipeline shown in Fig. 2.4, MeOH is injected at the wellhead so that more than 23 wt% MeOH is present in the free-water phase over the entire pipeline length.

As vaporized MeOH flows along the pipeline in Fig. 2.4, it dissolves into any produced brine or water condensed from the gas. Hydrate inhibition occurs in the free water, usually at accumulations with some change in geometry (e.g., a bend or pipeline dip along an ocean-floor depression) or some nucleation site (e.g., sand or weld slag).

Hydrate inhibition occurs in the aqueous liquid rather than in the vapor or condensate. While most of the MeOH dissolves in the water phase, a significant amount of MeOH either remains with the vapor or dissolves into any liquid hydrocarbon phase present, as calculated with the methods shown later in this section.

In Fig. 2.4, Notz[5] shows that the gas temperature increases from Mile 30 to Mile 45 with warmer (shallower) water conditions. From Mile 45 to Mile 50, however, a second cooling trend is observed as

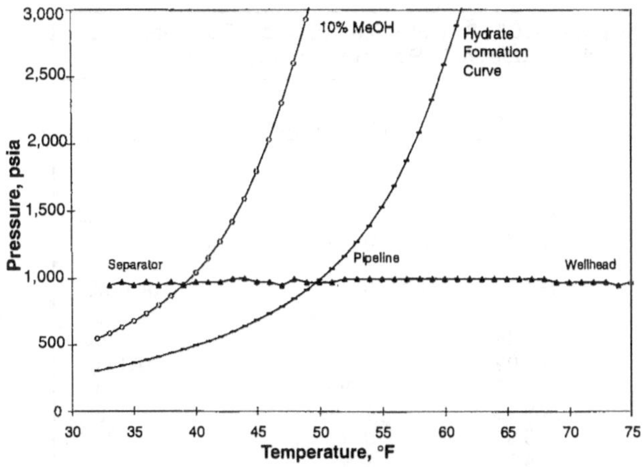

Fig. 2.5—Gas pipeline pressures and temperatures plotted on hydrate-formation curves. (Personal communication with J.L. Todd, Texaco, New Orleans (16 June 1997).

a result of a Joule-Thomson gas-expansion effect. MeOH exiting the pipeline in the vapor, aqueous, and condensate phases usually is not recovered because of the expense of regeneration.

Todd* provides an onshore pipeline simulation with a different behavior from that in Fig. 2.4. In Todd's simulations, typical gas-pipeline pressure drops were small relative to the overall pressure. This resulted in cooling at almost constant pressure, providing a straight, almost horizontal line between the pipeline endpoints on a plot like **Fig. 2.5.** Pipeline pressure drops are functions of several variables, and individual systems should be simulated for best results.

Example 2.3: Typical Offshore Platform Process. Manning and Thompson[6] detail a typical offshore platform process for a sweet crude oil with dissolved gas delivered to the platform at 1,000 psig and 120°F. **Fig. 2.6** shows the process, while **Table 2.2** gives process conditions and **Table 2.3** gives selected stream compositions.

*Personal communication with J.L. Todd, Texaco, New Orleans (18 June 1997).

The process was sized for production of 100,000 BOPD to the pipeline at the lease automatic custody transfer (LACT) unit, with 49 MMscf/D gas produced at 1,000 psig and an overall gas/oil ratio (GOR) of 491 scf/STB. The heavy ends of the crude are divided into five boiling-point cuts, and the mole fractions of individual gas components are given.

The three objectives of the platform process are (1) to separate the gas, water, and oil, providing an oil phase that has a very low vapor pressure and a water discharge to the ocean; (2) to dehydrate the gas to a water content at less than 7 lbm/MMscf before injection into the pipeline to shore; and (3) to compress the gas for transport to land.

Note that water separation and gas dehydration are vital for hydrate prevention, so that, even if the system cools into the hydrate pressure/temperature region that Fig. 2.5 shows, hydrate formation is prevented because of insufficient water. The export-pipeline-gas water content is below its water dewpoint (9 lbm/MMscf) at the lowest temperature (39°F), so free water does not condense from the gas phase.

The oil is stabilized by flow through a series of four separators, operating at 1,000, 300, 55, and 2 psig before the export-oil pipeline, so an oil-pipeline pressure greater than 17 psia prevents a gas phase. Hydrate formation is not a significant problem in the oil-export pipeline because relatively few hydrate formers (nitrogen, methane, ethane, propane, butanes, and CO_2) are dissolved in the oil and the water content is low.

The gas from each separator is compressed, cooled, and separated from liquid again before recombining the gas with the previous separator's gas for injection into the export gas line. The additional oil obtained after cooling the compressed gas amounts to about 1.5% of the total oil production.

In the process shown, 4,310-hp compressors represent the largest cost on the platform, with capital cost on the order of U.S. $800 to $1,500 (1990 dollars)/installed hp. These compressors are powered by fuel gas that operates at a low pressure (approximately 200 psig), usually fed from the inlet gas passing through a control valve with a substantial pressure reduction.

Pressure reductions at the fuel-gas takeoff cause cooling; therefore, that point is very susceptible to hydrate formation, particularly in winter months. Also, instrument gas lines require similar pressure reductions from a header. Todd *et al.*[7] observe that when fuel and/or instrument gas lines are blocked because of hydrates, the process frequently shuts down, resulting in pipeline cooling and significant hydrate blockages in the production line at restart.

Sec. 2.8 shows hydrate limits to pressure reductions through restrictions, such as valves and orifices.

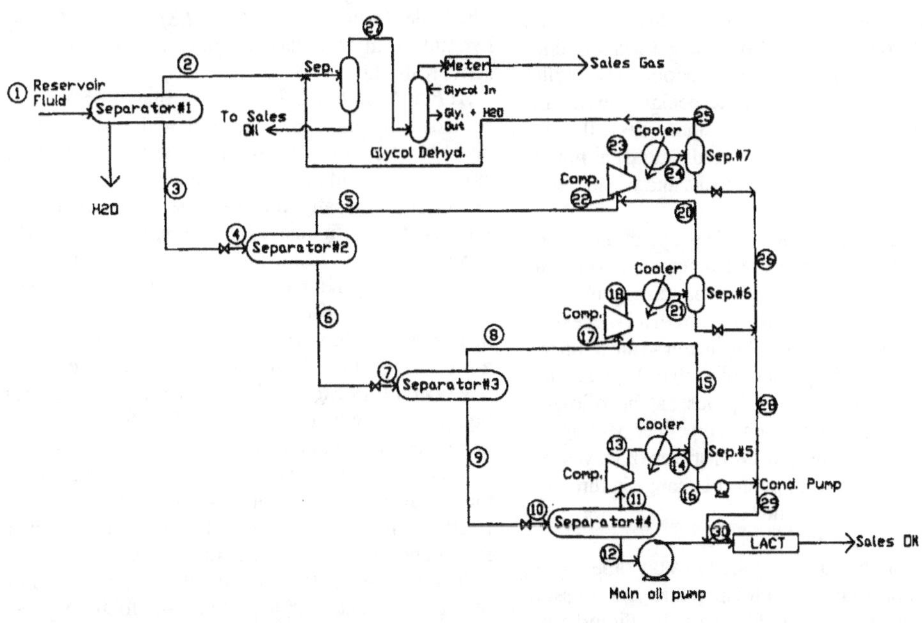

Fig. 2.6—Typical offshore platform schematic.[6] (Comp. = compressor; In. = injection; Sep. = separator). Source: F.S. Manning & R.E. Thompson, Oilfield Processing of Petroleum, Vol.I: Natural Gas Copyright PennWell Books, 1991.

TABLE 2.2—PROCESS CONDITIONS FOR TYPICAL OFFSHORE PLATFORM PROCESS IN FIG. 2.6.[6] (Source: F.S. Manning & R.E. Thompson, *Oilfield Processing of Petroleum*, VOL. I: Natural Gas Copyright Pennwell Books, 1991)

Location	Pressure (psia)	Temperature (°F)	(mol/hr)	Molecular Weight	Fraction Vapor	B/D at 60°F
1	1,019.7	120	1,2297.76	105.9	0.1821	0
2	1,019.7	120	2,238.98	18.79	1	0
3	1,019.7	120	10,058.78	125.29	0	111,807.9
4	3,14.7	115.86	10,058.78	125.29	0.2026	0
5	3,14.7	115.86	2,038.13	20.39	1	0
6	3,14.7	115.86	8,020.65	151.94	0	104,667.3
7	69.7	111.45	8,020.65	151.94	0.1084	0
8	69.7	111.45	869.66	27.44	1	0
9	69.7	111.45	7,150.99	167.09	0	101,141.7
10	16.7	106.22	7,150.99	167.09	0.0664	0
11	16.7	106.22	474.67	43.13	1	0
12	16.7	106.22	6,676.32	175.9	0	98,533.16
13	74.7	236.54	474.67	74.7	1	0
14	69.7	100	474.67	69.7	0.9464	0
15	69.7	100	449.21	69.7	1	0
16	69.7	100	25.47	69.7	0	199.99
17	69.7	106.27	1,318.87	32.2	1	0
18	3,19.7	280.91	1,318.87	32.2	1	0
19	3,14.7	100	1,318.87	32.2	0.8655	0
20	3,14.7	100	1,141.54	28.83	1	0
21	3,14.7	100	177.32	53.89	0	1,172.6
22	3,14.7	107.94	3,179.67	23.42	1	0
23	1,024.7	285.05	3,179.66	23.42	1	0
24	1,019.7	100	3,179.66	23.42	0.9926	0
25	1,019.7	100	3,156.23	23.27	1	0
26	1,019.7	100	23.43	43.18	0	144.6
27	1,019.7	104.9	5,395.21	21.41	1	0
28	3,14.7	95.43	200.75	52.64	0.0504	0
29	3,14.7	97.93	226.22	54.96	0.0275	0
30	3,14.7	104.75	6,902.53	171.93	0	100,000.1

Example 2.4: Onshore Gathering/Process System With Hydrate Occurrence Points. Fig. 2.7 shows a simple onshore gas gathering/processing system. The gas comes from the well at Point A and then may be heated before Point B, where it passes through a pressure-regulation valve to Point C. At Point C, depending on the climate and gas conditions, the operator may elect to dehydrate or to inject inhibitor to prevent hydrate formation at Point D, before the gas enters the gathering system. Gases from several wells may be gathered in this manner before they are of sufficient quantity to process economically.

Gas from the gathering line proceeds some distance to the processing plant, where the entire stream may be dehydrated or inhibitor may be injected (E) before compression (F), cooling in a chiller (G), and expansion, through either a valve or a turboexpander, to provide liquid recovery from the gas stream. The gas then goes to the user where it is burned for heating value. The recovered natural gas liquids (NGL's) typically are used as chemical feedstocks and liquid petroleum gas (LPG) fuel.

In such a process, hydrate blockage jeopardizes operation at each point that has conditions of high pressure, low temperature, or cooling by expansion, such as across the various valves or across a turboexpander. Conversely, if hydrate formation has occurred, it is necessary to decrease the pressure at each end of the plug, increase the temperature, and/or inject an inhibitor at the point of blockage. Typically hydrate formation and accumulation occurs in a process at the following points: (1) low-lying equipment points, such as pipelines passing under a roadway; (2) points of gas expansion, downstream of valves, expanders, and other such equipment; (3) points of change in direction of flow, such as pipe elbows; and (4) points of flow obstruction, such as screens preceding heat exchangers.

As an example, to prevent hydrate formation in a pipeline during cold weather, it is common to maintain temperature and pressure conditions in the hydrate-free zone within the vapor/liquid two-phase region on the right of Figs. 2.4 and 2.5. In addition to pipeline-burial, a common technique is to install a heater system at the wellhead (between Points A and B in Fig. 2.7) to raise the gas temperature.[9] The heat exchanger can be as simple as a barrel of hot oil through which a coiled gas pipe passes.* Such a heating system may be a cost-effective alternative to inhibitor injection, particularly in cases where cold weather is of short duration. However, hydrates may re-form downstream if the pipeline is long enough for heat losses to cool the gas below the hydrate temperature.

Handwerk* provides the following three illustrations at which hydrate formation was typical in a system like that shown in Fig. 2.7.

At Point J in Fig. 2.7. Gas that has not been dehydrated (i.e., wet gas) is taken from a high-pressure well or gathering line and used for heater or compressor-engine fuel at lower pressures. The gas may pass through two control valves (pressure reductions) in series. The Joule-Thomson effect (Sec. 2.8) reduces the temperature so that hydrates form. Hydrates plug the piping downstream of the valve at some restriction.

At Point F in Fig. 2.7. Cryogenic gas-processing units are usually preceded by molecular-sieve dehydration systems. Over time, enough water can pass through these to allow hydrates to collect slowly in the cold sections of downstream equipment to cause flow reduction. Typically this becomes a problem at inlet dust/dirt screens at plate/fin heat exchangers, valves, or turboexpanders.

Such problems occur infrequently (at intervals of some weeks). However, if regeneration of the desiccant is not adequate or if the dehydration unit is not operating properly, hydrates can plug equipment in a matter of hours. The usual on-line short-term remedy is to inject a small amount of MeOH either periodically or continuously.

At Point G in Fig. 2.7. Wet gas being processed for hydrocarbon-liquid recovery by refrigeration often forms hydrates in the chiller or related cold-gas/warm-gas heat exchangers. The wet inlet gas flows through the small channels of the exchangers or chillers and the metal wall temperature is cold enough to cause hydrates to freeze on the surface. Hydrate fouling causes exchanger heat-transfer reduction and a high pressure drop.

In some cases, ethylene glycol is sprayed continuously into the feed gas to exchangers to prevent—or to dissociate—hydrate accumulation. Hydrate problems that occur in ethylene glycol inhibition systems normally result from three factors.

*Personal communication with G.E. Handwerk, Golden, Colorado (3 August 1988).

TABLE 2.3—GAS AND LIQUID COMPOSITION ON PLATFORM.[6] (Source: F.S. Manning & R.E. Thompson, *Oilfield Processing of Petroleum*, VOL. I: Natural Gas Copyright Pennwell Books, 1991)

	1	2	3	5	6	8
Component Mole Fraction	Inlet Fluid	Gas Out First Separator	Liquid Out First Separator	Gas Out Second Separator	Liquid Out Second Separator	Gas Out Third Separator
Nitrogen	0.0078	0.0287	0.0031	0.0137	0.0005	0.0040
CO_2	0.0005	0.0009	0.0004	0.0012	0.0002	0.0015
Methane	0.3386	0.8705	0.2202	0.8074	0.0710	0.5605
Ethane	0.0563	0.0607	0.0553	0.1060	0.0424	0.2118
Propane	0.0440	0.0213	0.0491	0.0416	0.0510	0.1232
i-butane	0.0121	0.0033	0.0140	0.0062	0.0160	0.0203
n-butane	0.0342	0.0073	0.0402	0.0133	0.0470	0.0444
i-pentane	0.0185	0.0022	0.0221	0.0036	0.0269	0.0118
n-pentane	0.0244	0.0023	0.0293	0.0036	0.0359	0.0120
Hexane	0.0429	0.0018	0.0520	0.0024	0.0647	0.0075
248°F	0.0996	0.0009	0.1216	0.0010	0.1522	0.0027
340°F	0.0714	0.0001	0.0873	0.0001	0.1094	0.0003
413°F	0.0611	0.0000	0.0747	0.0000	0.0937	0.0000
472°F	0.0544	0.0000	0.0665	0.0000	0.0834	0.0000
657°F	0.1342	0.0000	0.1641	0.0000	0.2058	0.0000
Total mol/hr	12,297.75	2,238.98	10,058.78	2,038.13	8,020.67	869.66

	9	11	12	14	15	16
Component Mole Fraction	Liquid Out Third Separator	Gas Out Third Separator	Liquid Out Fourth Separator	Fifth Separator Inlet	Gas Out Sixth Separator	Liquid Out Sixth Separator
Nitrogen	0.0000	0.0004	0.0000	0.0004	0.0005	0.0000
CO_2	0.0001	0.0009	0.0000	0.0009	0.0009	0.0000
Methane	0.0115	0.1615	0.0008	0.1615	0.1704	0.0043
Ethane	0.0219	0.2399	0.0063	0.2399	0.2517	0.0318
Propane	0.0422	0.2789	0.0253	0.2789	0.2880	0.1190
i-butane	0.0155	0.0597	0.0124	0.0597	0.0598	0.0562
n-butane	0.0474	0.1393	0.0408	0.1393	0.1371	0.1783
i-pentane	0.0287	0.0407	0.0278	0.0407	0.0368	0.1108
n-pentane	0.0388	0.0418	0.0385	0.0418	0.0360	0.1438
Hexane	0.0716	0.0267	0.0748	0.0267	0.0169	0.1995
248°F	0.1704	0.0092	0.1819	0.0092	0.0018	0.1398
340°F	0.1227	0.0008	0.1313	0.0008	0.0000	0.0145
413°F	0.1051	0.0001	0.1125	0.0001	0.0000	0.0020
472°F	0.0935	0.0000	0.1002	0.0000	0.0000	0.0000
657°F	0.2308	0.0000	0.2472	0.0000	0.0000	0.0000
Total mol/hr	7,150.98	474.66	6,676.31	474.66	449.2	25.46

	17	20	21	23	25	26
Component Mole Fraction	Sixth Separator Inlet	Gas Out Sixth Separator	Liquid Out Sixth Separator	Seventh Separator Inlet	Gas Out Seventh Separator	Liquid Out Seventh Separator
Nitrogen	0.002783	0.000467	0.000169	0.0000[illegible]	0.00999	0.002135
CO_2	0.001304	0.000935	0.000395	0.00128	0.001283	0.000854
Methane	0.42762	0.170392	0.061975	0.69145	0.694509	0.279249
Ethane	0.225381	0.251714	0.125021	0.154435	0.154317	0.170367
Propane	0.179342	0.288001	0.248351	0.08717	0.086334	0.199829
i-butane	0.033794	0.05984	0.081205	0.013479	0.013199	0.051238
n-butane	0.075951	0.137066	0.218463	0.027843	0.027092	0.12895
i-pentane	0.020328	0.036754	0.086336	0.005897	0.005605	0.04526
n-pentane	0.020161	0.03602	0.094344	0.005419	0.005098	0.048676
Hexane	0.010736	0.016941	0.065133	0.002365	0.002091	0.039283
248°F	0.002404	0.001848	0.017143	0.000654	0.000456	0.027327
340°F	0.000174	2.23×10^{-5}	0.001297	6.6×10^{-5}	2.53×10^{-5}	0.005551
413°F	2.27×10^{-5}	0	0.000169	9.44×10^{-6}	0	0.001281
472°F	0	0	0	0	0	0
657°F	0	0	0	0	0	0
Total mol/hr	1,318.88	449.2	177.33	3,179.65	3,156.23	23.42

- Unequal distribution of glycol in the heat exchanger tubes.
- Excessive glycol viscosity so that flow is restricted through the tubes.
- Inadequate glycol injection for inhibition.

With the preceding overview of a typical process and examples of places where hydrate formation occurs in both on- and offshore processes, the remainder of this chapter considers hydrate prediction and prevention techniques.

2.5 One-Minute Estimate of Hydrate-Formation Conditions (Accurate to 50%)

Assuming the pipeline pressure drop to be relatively small, the engineer may do a rough estimation to determine whether the pipeline will operate in the hydrate region. As a first approximation, the engineer should first calculate the pressure at which hydrates form in the coolest portion of the system (e.g., at the lowest deep-ocean temperature or 38 to 40°F) so that, if the pipeline pressure is greater, inhibition should be considered in the pipeline design and operation. Such an approximation may indicate the need for more accurate calculations to determine the amount of inhibition required.

TABLE 2.3—GAS AND LIQUID COMPOSTION ON PLATFORM (Continued).[6] (Source: F.S. Manning & R.E. Thompson, *Oilfield Processing of Petroleum*, VOL. I: Natural Gas Copyright Pennwell Books, 1991)

	27	28	29	30
Component Mole Fraction	Sales Gas	Liquid Line	Liquid Line	Sales Oil
Nitrogen	0.017764	0.000398	0.000354	1.3×10^{-5}
CO_2	0.00111	0.000448	0.000398	2.32×10^{-5}
Methane	0.767528	0.087314	0.077977	0.003338
Ethane	0.115474	0.130298	0.119176	0.010048
Propane	0.059332	0.242716	0.22876	0.032016
i-butane	0.009112	0.077701	0.075325	0.014435
n-butane	0.018863	0.207999	0.204668	0.046189
i-pentane	0.004178	0.081536	0.084829	0.029695
n-pentane	0.003929	0.089057	0.095217	0.040401
Hexane	0.00197	0.062111	0.077535	0.074892
248°F	0.000649	0.018329	0.032004	0.176941
340°F	7.41×10^{-5}	0.001793	0.003227	0.12715
413°F	1.3×10^{-5}	0.000249	0.000442	0.108848
472°F	3.71×10^{-6}	4.98×10^{-5}	8.84×10^{-5}	0.096918
657°F	0	0	0	0.239094
Total mol/hr	5,395.22	200.77	226.22	6,902.57

2.5.1 Rules of Thumb. In this book, rules of thumb are printed in bold type. These rules of thumb are based on experience, and they are intended as engineering guides for further action. For example, using a rule of thumb, an engineer might determine that a more accurate calculation is needed for inhibitor injection amounts or that further consideration of hydrates is unnecessary. These rules of thumb are not intended to be absolute truths, and exceptions always can be found. Where possible, the accuracy of each rule of thumb is provided.

The first rule of thumb describes the conditions for hydrate formation at a deep-ocean-bottom temperature of 39°F. This temperature is nominally the lowest temperature on the deep ocean bottom; exceptions result from cold subsea currents. For convenience, 39°F also serves to approximate the lowest onshore pipeline temperature during winter months. While onshore pipeline temperatures can be warmer or cooler than 39°F, slightly lower temperatures (below 32°F) normally are avoided when free water is in the line because ice is a more hazardous flow obstruction than hydrates. For onshore pipelines with temperatures different from 39°F, the hydrate prediction methods of Secs. 2.6 and 2.7 are recommended.

Rule-of-Thumb 1. At 39°F, hydrates will form in a natural gas system if free water is available and the pressure is greater than 166 psig.

Hydrate-formation data at 39°F were averaged for 20 natural gases with an average formation pressure of 181 psia.[8] Of the 20 gases, the lowest formation pressure was 100 psig for a gas with 7 mol% C_3H_8, while the highest value was 300 psig for a gas with 1.8 mol% C_3H_8.

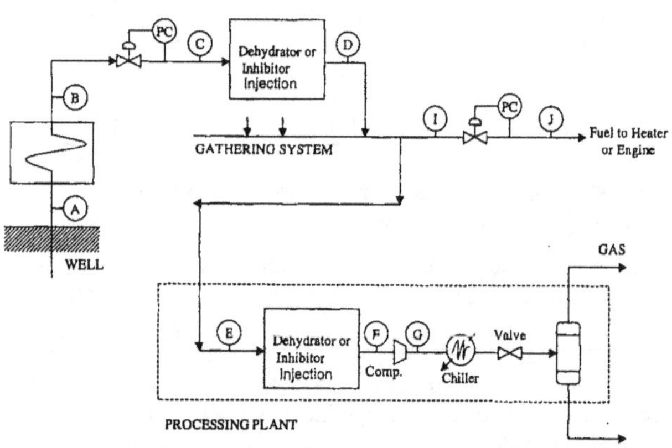

Fig. 2.7—Likely points of hydrate formation in onshore process.[8] (Comp.=compressor)

Rule-of-Thumb 1 indicates that most onshore and offshore pipeline pressures greatly exceed the hydrate-formation condition, indicating the following:

• Gas drying or inhibition is necessary for temperatures approaching 39°F.

• A more accurate estimation procedure should be considered.

• Hydrate-formation pressures depend on the gas composition and are particularly sensitive to the amount of propane present.

It should be reiterated that hydrates can form at temperatures in excess of 39°F when the pressure is elevated, as in the case of warmer temperatures in shallower water. More accurate estimations of hydrate-formation conditions over a broad temperature range are made by the method in the following section.

2.6 Ten-Minute Estimation of Hydrate Formation/Inhibition (Accurate to 25%)

As a second approximation of hydrate-formation conditions, the design/facilities engineer can perform two calculations.

1. A pipeline pressure/temperature flow simulation should be done to determine the conditions between the wellhead and the pipeline discharge.

2. Hydrate-formation conditions like those shown in Fig. 2.4 should be calculated, and pressures and temperatures of vapor and aqueous liquid inhibited by various amounts (including 0 wt%) of MeOH or MEG should be considered.

The intersection of these two lines determines the pressure and temperature at which hydrates form in a pipeline. As Figs. 2.4 and 2.5 (Example 2.2) show, it is very likely that a deep offshore pipeline or an onshore pipeline in winter have hydrate-formation conditions when free water is present. The engineer then specifies the amount of inhibitor needed to keep the entire pipeline in the fluid region without hydrate formation.

Step 1 in this calculation—the flow simulation of the pipeline—is beyond the scope of this book and should be considered as a separate, prerequisite problem, perhaps done by the engineering staff at the home office. As an alternative, if a pipe-flow simulation is not readily available, the engineer may wish to assume that the contents of a long pipeline eventually come to the ambient or ocean-bottom temperature at the inlet pipeline pressure to obtain the worst case scenario for hydrate formation.

Step 2, enabling estimations of hydrate-formation pressures and temperatures, is one of the principal goals of this book, as discussed in this and in the following section. The next methods (Secs. 2.6 and 2.7) may then be used directly to determine the amount of MeOH or MEG needed to prevent hydrate formation at those conditions.

2.6.1. Hydrate-Formation Conditions by the Gas-Gravity Method. The simplest method to determine the hydrate-formation

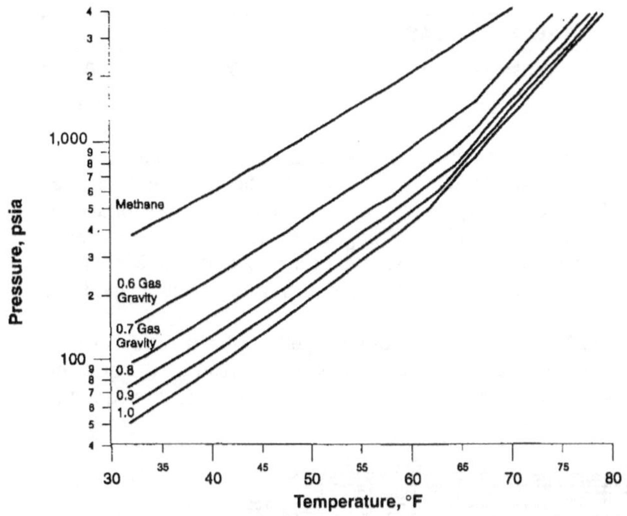

Fig. 2.8—Hydrate-formation curves for various gas gravities.[10]

temperature and pressure is by means of gas gravity, defined as the molecular weight of the gas divided by that of air. **Fig. 2.8** is a chart of hydrate-formation curves for different gas gravities. To use this chart, the gas gravity is calculated and the lowest temperature of the pipeline/process is specified. The pressure at which hydrates form is read directly from the chart at the gas gravity and temperature.

To the left of every line, hydrates form with a gas of that gravity, while for pressures and temperatures to the right of the line, the system is hydrate-free The following example from the original work by Katz[11] illustrates chart use.

Example 2.5: Calculating Hydrate-Formation Conditions With the Gas-Gravity Chart. Find the pressure at which a gas composed of 92.67 mol% methane, 5.29% ethane, 1.38% propane, 0.182% *i*-butane, 0.338% *n*-butane, and 0.14% pentane forms hydrates with free water at 50°F.

Solution. The gas gravity is calculated as 0.603 by the procedure in **Table 2.4** and the following gas-gravity equation.

$$\gamma_g = \frac{M_g}{M_a} = \frac{17.47}{28.96} = 0.603, \quad \ldots\ldots\ldots\ldots\ldots\ldots \quad (2.1)$$

where γ_g = gas gravity, M_g = gas molecular weight, and M_a = air molecular weight.

At 50°F in Fig. 2.8, the hydrate pressure is read as 450 psia at a gas gravity of 0.603.

The user is cautioned that this method is only approximate for several reasons. Fig. 2.8 was generated for gases containing only hydrocarbons and so should be used with caution for those gases with substantial amounts of CO_2, H_2S, or N_2. In addition, the estimated inaccuracies for the hydrate equilibrium temperature, T_{eq}, and pressure, p_{eq}, are maximized for 0.6 gravity gas ±7°F or ±500 psig.[12] In the 50 years since the generation of this chart, more hydrate data and prediction methods have caused the gravity method to be used as a first estimate whose principal asset is ease of calculation. Sec. 2.7 provides one of the most accurate methods for calculation of hydrate conditions, but it requires additional time and a computer.

TABLE 2.4—GAS GRAVITY CHART

Component	Mole Fraction y_i	Molecular Weight M	Fraction Molecular Weight in Mixture y_iM
Methane	0.9267	16.043	14.867
Ethane	0.0529	30.070	1.591
Propane	0.0138	44.097	0.609
i-butane	0.00182	58.124	0.106
n-butane	0.00338	58.124	0.196
Pentane	0.0014	72.151	0.101
Total	1.000	Average Molecular Weight = 17.470	

2.6.2. Estimating the Hydrate Inhibitor Needed in the Free-Water Phase.

The gas-gravity chart in Table 2.4 may be combined with the Hammerschmidt equation to estimate the hydrate depression temperature for several inhibitors in the aqueous liquid:

$$\Delta T = \frac{C_I W_I}{M_I(100 - W_I)}, \quad \ldots\ldots\ldots\ldots\ldots\ldots \quad (2.2)$$

where ΔT = hydrate depression, $(T_{eq} - T_{op})$, °F at the pressure, C_I = constant for a particular inhibitor (2,335 for MeOH; 2,000 for MEG), W_I = weight per cent of the inhibitor in the liquid, and M_I = molecular weight of MeOH (32) or MEG (62).

The Hammerschmidt equation was generated in 1934 and has been used to determine the amount of inhibitor needed to prevent hydrate formation, as indicated in Example 2.6.[13] The equation was based on more than 100 natural-gas-hydrate measurements with inhibitor concentrations of 5 to 25 wt% in water. The accuracy of the Hammerschmidt equation is surprisingly good; tested against 75 data points, the average error in ΔT was 5%.

For higher MeOH concentrations (up to 87 wt%) the temperature depression caused by MeOH can be calculated by a modification of Eq. 2.2, where x_{MeOH} is mole fraction MeOH in aqueous phase[14]

$$\Delta T = -129.6 \ln(1 - x_{MeOH}). \quad \ldots\ldots\ldots\ldots\ldots\ldots \quad (2.3)$$

Example 2.6: MeOH Concentration With the Hammerschmidt Equation. Estimate the MeOH concentration needed to provide hydrate inhibition at 450 psia and a temperature of 39°F for a gas composed of 92.67 mol% methane, 5.29% ethane, 1.38% propane, 0.182% *i*-butane, 0.338% *n*-butane, and 0.14% pentane.

Solution. The gas is the same composition and pressure as that in Example 2.5, with the gas gravity previously determined to be 0.603 and uninhibited hydrate-formation conditions of 50°F and 450 psia. Inhibition is required because the pipeline operates at 39°F and 450 psia, well within the hydrate-formation region. The weight percent of inhibitor needed in water phase is determined with the Hammerschmidt Eq. 2.1, with the following values: ΔT = temperature depression (50°F − 39°F = 11°F), M_{MeOH} = molecular weight for MeOH (= 32), C_I = constant for MeOH (= 2335), and W_I = weight percent inhibitor.

Rearranging Eq. 2.2,

$$W_I = \frac{100 M_{MeOH} \Delta T}{M_{MeOH} \Delta T + C_I} = \frac{100 \times 32 \times 11}{32 \times 11 + 2335} = 13.1.$$

The MeOH in the water phase is predicted as 13.1 wt% to provide hydrate inhibition at 450 psia and 39°F for this gas. The engineer may wish to provide an operational safety factor by adding more MeOH.

2.6.3. Amount of Inhibitor Injected Into Pipeline.

While the Hammerschmidt equation enables estimation of the weight percent MeOH (or MEG) needed in the free-water phase, three other quantities are necessary to estimate the amount of inhibitor injected into the pipeline: (1) the amount of the free water phase, (2) the amount of inhibitor lost to the gas phase, and (3) the amount of inhibitor lost to the condensate phase.

The amount of the free-water phase is multiplied by the inhibitor concentration from the Hammerschmidt equation, just as inhibitor concentrations in the gas and condensate are multiplied by the flows of the vapor and condensate. Because hydrate inhibition occurs in the water phase, inhibitor concentrations in the gas and condensate phases are usually counted as economic losses. MeOH recovery typically is too expensive either at onshore locations or on platforms.

Amount of Water Phase. The water phase has two sources: produced water and water condensed from the hydrocarbon phases. The amount of produced water can be determined only from data from the well, and water production tends to increase over the well's lifetime.

Water condensed from the hydrocarbon phases may be calculated. The water content of condensates usually is negligible, but water condensed from gases can be substantial. The amount of water condensed is the difference in the inlet- and outlet-gas water content multiplied by the gas flow rate.

Rule-of-Thumb 2. For long pipelines approaching the ocean-bottom temperature of 39°F, Table 2.5 gives the lowest water content of the outlet gas.

TABLE 2.5—GAS WATER CONTENT AT 39°F

Pipe pressure, psia	500	1,000	1,500	2,000
Water content, lbm/MMscf	15.0	9.0	7.0	5.5

An inlet-gas water-content analysis is used, if available. Then the water content of the outlet gas (Rule-of-Thumb 2) may be subtracted from the inlet gas content to determine the water condensed per million standard cubic feet of gas. When an inlet-gas water content is not available, a water-content chart such as **Fig. 2.9** may be used to obtain the water content of both the inlet and outlet gas from the pipeline.

In Fig. 2.9 the temperature of the pipeline inlet or outlet is found on the abscissa (x axis) and water content is read on the ordinate (y axis) at each pipeline pressure. The engineer is cautioned not to

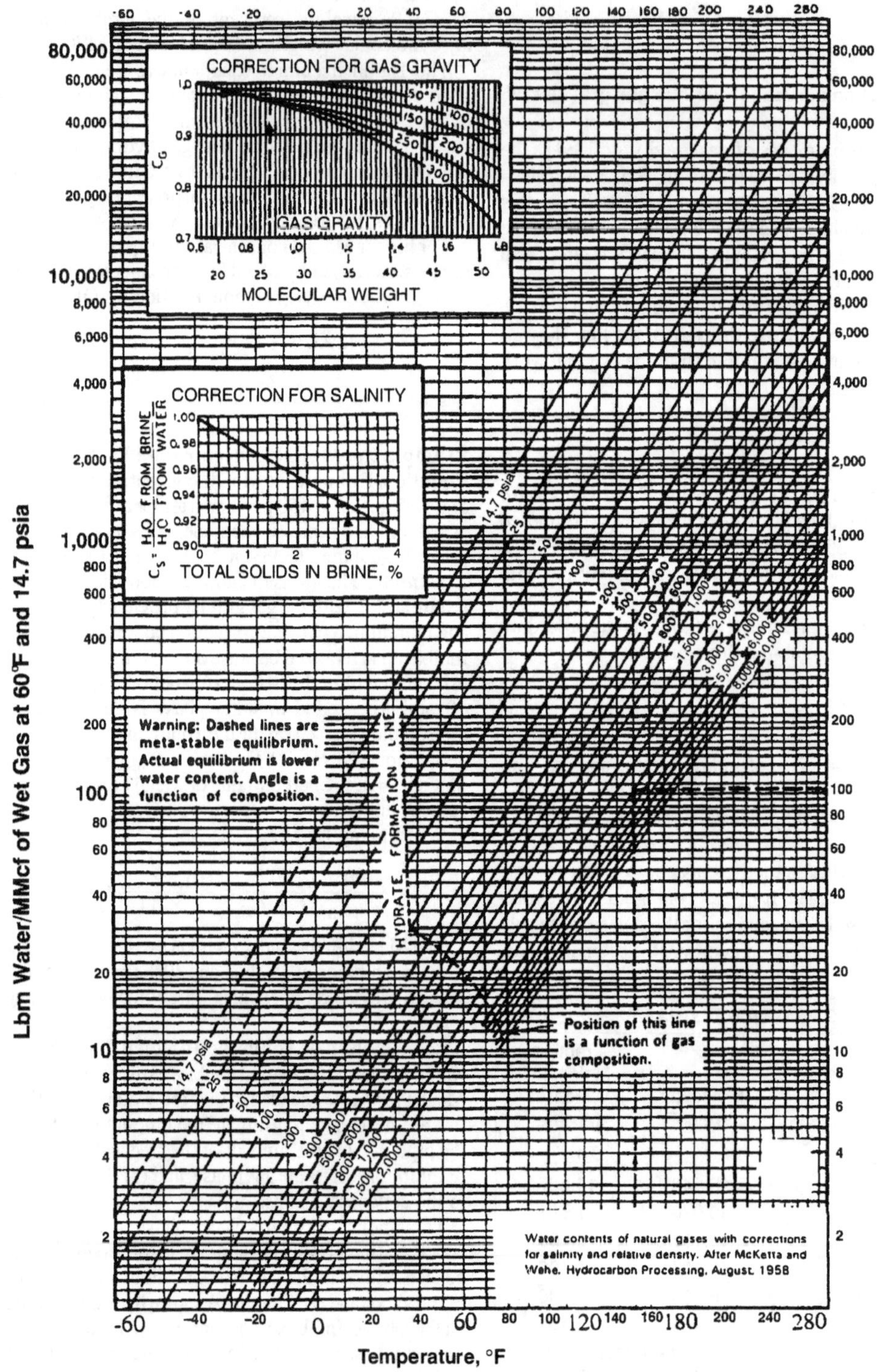

Fig. 2.9—Water formation curve.[15]

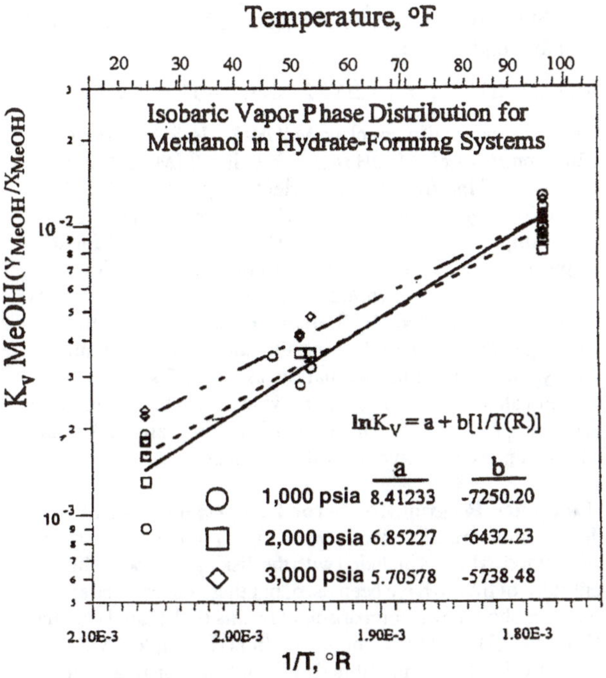

Fig. 2.10—MeOH lost to vapor.[8]

use the water-content chart at temperatures significantly below 38°F. At lower temperatures, the actual water content deviates from the line because of hydrate formation. Step 3 of Example 2.7 illustrates condensed-water calculation with Fig. 2.9.

Amount of Inhibitor Lost to the Gas Phase. The Hammerschmidt equation provides only the amount of MeOH needed in the free-water phase at the point of hydrate inhibition, while two other phases represent potential losses of MeOH. The amount of MeOH or MEG loss into the gas phase also should be considered with the following rules of thumb.

Rule-of-Thumb 3. At 39°F and pressures greater than 1,000 psia, the MeOH lost to the vapor phase is 1 lbm MeOH/MMscf for every weight percent MeOH in the free-water phase.

Rule-of-Thumb 4. At 39°F and pressures greater than 1,000 psia, the maximum amount of MEG lost to the gas is 0.02 lbm/MMscf.

The MeOH loss chart in **Fig. 2.10** shows that, at typical pipeline conditions, the amount of MeOH in the vapor may be 0.3 mol% of that in the water phase. Rule-of-Thumb 3 is valid except for low-water amounts, when the MeOH vapor loss can be substantially higher and the method in Sec. 2.7.3 should be used. **Fig. 2.11** validates Rule-of-Thumb 4 for MEG. Note that the data for Fig. 2.10 were obtained in 1995 for the mole fraction ratio of inhibitor in the vapor over that in the aqueous phase; to use the figure, the water-phase weight percent inhibitor must be converted to mole percent. Example 2.7 illustrates MeOH loss to the gas phase.

Amount of Inhibitor Lost to the Liquid Phase. Two general rules of thumb can be applied to inhibitor losses in the condensate.

Rule-of-Thumb 5. MeOH concentration dissolved in condensate is 0.5 wt%.

Rule-of-Thumb 6. The mole fraction of MEG in a liquid hydrocarbon at 39°F is 0.03% of the water-phase mole fraction of MEG.

Even with low losses of MEG relative to MeOH in both the gas and the liquid, it is important to remember that MeOH is a much more effective inhibitor than ethylene glycol on a weight basis. The predominance of use of MeOH is because of this effectiveness and because MeOH flows easily to the point of hydrate formation.

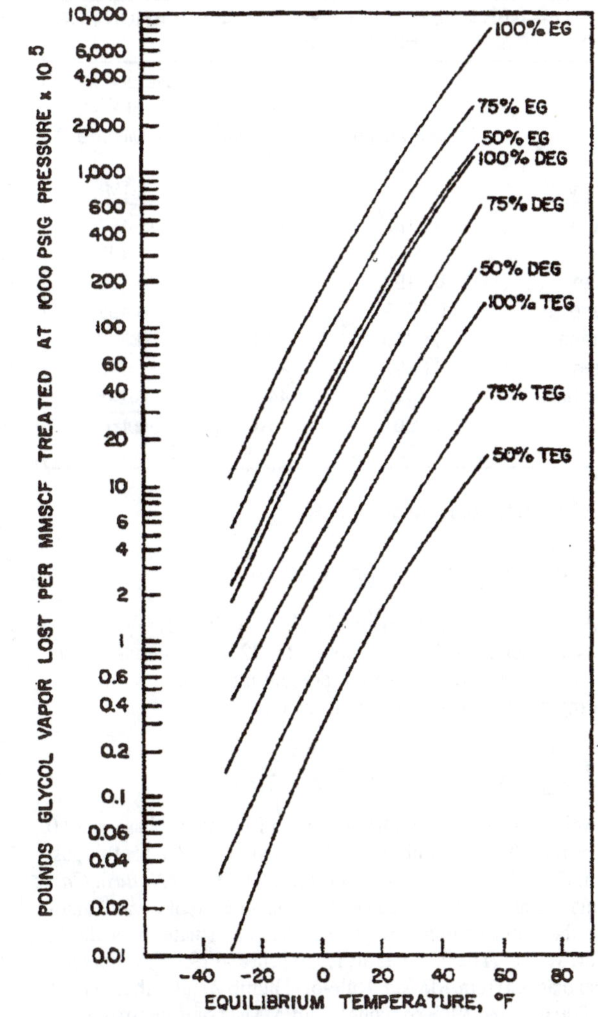

Fig. 2.11—MEG lost to vapor.[16] (EG=MEG; DEG=diethylene glycol; TEG=triethylene glycol).

2.6.4 Example Calculation of Amount MeOH Injection. The next sample calculation uses all the concepts presented in Sec. 2.6.

Example 2.7: MeOH Injection Rate. A pipeline with the following gas composition has inlet pipeline conditions of 195°F and 1,050 psia. The gas flowing through the pipeline is cooled by the surrounding water to a temperature of 38°F. The gas also experiences a pressure drop to 950 psia. Gas exits the pipeline at a rate of 3.2 MMscf/D. The pipeline produces condensate at a rate of 25 B/D, with an average density of 300 lbm/bbl and an average molecular weight of 90 lbm/lbm mol. Produced free water enters the pipeline at a rate of 0.25 B/D.

Find the rate of MeOH injection needed to prevent hydrates in the pipeline for a natural gas composition (mol%): methane = 71.60%, ethane = 4.73%, propane=1.94%, *n*-butane = 0.79%, n-pentane = 0.79%, carbon dioxide = 14.19%, nitrogen = 5.96%.

Solution. The basis for these calculations was chosen as 1 MMscf/D.

*Step 1. Calculate Hydrate-Formation Conditions With the Gas-Gravity Chart in **Table 2.6** and Eq. 2.1.*

$$\gamma_g = \frac{M_g}{M_a} = \frac{22.708}{28.966} = 0.784.$$

Reading the gas-gravity chart (Fig. 2.8), the hydrate temperature is 65°F at 1,000 psia.

Step 2. Calculate the Weight Percent MeOH Needed in the Free-Water Phase.

TABLE 2.6—GAS MOLECULAR WEIGHT CALCULATION FOR EXAMPLE 2.7

Component	Mole Fraction y_i	Molecular Weight M	Average Molecular Weight in Mixture y_iM
Methane	0.7160	16.04	11.487
Ethane	0.0473	30.07	1.422
Propane	0.0194	44.09	0.855
n-butane	0.0079	58.12	0.459
n-pentane	0.0079	72.15	0.570
Nitrogen	0.0596	28.01	1.670
Carbon dioxide	0.1419	044.01	6.245
	1.000	Average Molecular Weight = 22.708	

The Hammerschmidt equation is

$$\Delta T = \frac{C_I W_I}{100 M_{\text{MeOH}} - M_{\text{MeOH}} W_I},$$

where ΔT = temperature depression (65°F − 38°F = 27°F), M_{MeOH} = molecular weight for MeOH (= 32.0), C_I = constant for MeOH (−2,335), and W_I – weight percent inhibitor in water.

Rearranging the Hammerschmidt equation

$$W_I = \frac{100 M_{\text{MeOH}} \Delta T}{M_{\text{MeOH}} \Delta T + C_I} = \frac{100 \times 32 \times 27}{32 \times 27 + 2,335} = 27.$$

The weight percent of MeOH needed in free-water phase is 27.0 wt% to provide hydrate inhibition at 1,000 psia and 38°F for this gas.

Step 3. Calculate the Mass of Liquid H_2O/MMscf of Natural Gas. 1. Calculate mass of condensed H_2O. In the absence of a water analysis, use the water-content chart (Fig. 2.9) to calculate the water in the vapor/MMscf. The inlet gas (at 1,050 psia and 195°F) water content is read as 600 lbm/MMscf. Rule-of-Thumb 2 states that exiting gas at 1,000 psia and 39°F contains 9 lbm/MMscf of water in the gas. The mass of liquid water caused by condensation is

$$\frac{600 \text{ lbm}}{\text{MMscf}} - \frac{9 \text{ lbm}}{\text{MMscf}} = \frac{591 \text{ lbm}}{\text{MMscf}}.$$

2. Calculate mass of produced H_2O flowing into the line. Convert the produced water of 0.25 B/D to 27.4 lbm/MMscf:

$$(0.25 \text{ B/D } H_2O)\left(\frac{42 \text{ gal}}{\text{bbl}}\right)\left(\frac{8.34 \text{ lbm}}{32 \text{ MMscf}}\right) = 27.4 \frac{\text{lbm } H_2O}{\text{MMscf}}.$$

Total mass of water/MMscf gas. Add the condensed and produced water

$$\frac{591 \text{ lbm}}{\text{MMscf}} + \frac{27.4 \text{ lbm}}{\text{MMscf}} = \frac{618.4 \text{ lbm}}{\text{MMscf}}.$$

Step 4. Calculate the Rate of MeOH Injection. MeOH exists in three phases: water, gas, and condensate. The total mass of MeOH injected into the gas is calculated as follows:

1. Calculate mass of MeOH, m_{MeOH}, in the water phase in pounds. 27.0 wt% MeOH is required to inhibit the free-water phase, and the mass of water/MMscf was calculated at 618.4 lbm. The mass of MeOH in the free-water phase per MMscf is

$$27 \text{ wt\%} = \frac{m_{\text{MeOH}}}{m_{\text{MeOH}} + 618 \text{ lbm } H_2O} \times 100\%.$$

Solving for m_{MeOH} gives 228.7 lbm MeOH in the water phase.

2. Calculate m_{MeOH} lost to the gas. Rule-of-Thumb 3 states that the maximum amount of MeOH lost to the vapor phase is 1 lbm MeOH/MMscf for every wt% MeOH in the water phase. Because there is 27 wt% MeOH in the water, the maximum amount of MeOH lost to the gas is 27 lbm/MMscf.

3. Calculate m_{MeOH} lost to the condensate. Rule-of-Thumb 5 states that the MeOH concentration in the condensate is 0.5 wt%. Because 1 bbl of hydrocarbon weighs about 300 lbm, the amount of MeOH in the condensate is

$$0.005(300 \text{ lbm/bbl})(25 \text{ B/D})(1\text{D}/3.2 \text{ MMscf}) = 11.7 \text{ lbm/MMscf}.$$

4. Calculate the total amount of MeOH/MMscf. MeOH in water = 228.7 lbm/MMscf; MeOH in gas = 27 lbm/MMscf; MeOH in condensate = 11.7 lbm/MMscf; total MeOH injection = 267.4 lbm/MMscf (or 40.33 gal/MMscf at a MeOH density of 6.63 lbm/gal).

In this example, the amount of MeOH lost to the gas and condensate is approximately 11% of the total amount injected. However, with large amounts of condensate, it is not uncommon to have as much as 90% of the injected MeOH dissolved in the condensate (primarily) and gas phases. In such cases, the rules of thumb should be replaced by a more accurate calculation, as Sec. 2.7 shows.

The hand-calculation example is provided for understanding of the second approximation. The spreadsheet program makes the method much more convenient for the engineer.

2.6.5 Computer Program for Second Approximation. Shuler* provides a computerized version (HYDCALC) of the previous calculation method, which is included with the disk in this book. Slightly different rules of thumb have been used, but these differences are insignificant, as shown by a comparison of results of the hand calculation (Example 2.7) with the computer method (Example 2.8).

HYDCALC is a PC-compatible spreadsheet that provides an initial estimate of pipeline MeOH injection for hydrate inhibition. To use HYDCALC, obtain access to a Microsoft Excel Ver. 7.0 spreadsheet program and copy HYDCALC into a hard drive directory. Start Excel Ver. 7.0 and open the file HYDCALC.

Once the file is opened, the user will see text in three different colors on a color screen—black, red, and blue. The red text signifies required user inputs composed of the following eight pieces of information to start the program.

1. Pipeline inlet pressure—starting high pressure.
2. Cold-pipeline pressure—pressure at the coldest part of the pipeline.
3. Pipeline-inlet temperature—starting warm temperature.
4. Cold-pipeline temperature—temperature at the coldest part of the pipeline.
5. Gas gravity—gas gravity, calculated by the steps in Sec. 2.6.1 and Example 2.5.
6. Gas flow rate—gas flow in the pipeline measured in MMscf/D.
7. Condensate rate—condensate flow in the pipeline measured in B/D.
8. Formation water rate—produced water flowing into the pipeline in B/D.

Once these values are entered, HYDCALC displays calculations for both intermediate results (in black) and the amount of MeOH or glycol to be injected (in blue). Everything in the next example is printed in black. Example 2.8 provides a prescription for the use of this method.

Example 2.8: Use of HYDCALC To Find Amount of MeOH and Glycol Injection. This spreadsheet problem is identical to the problem worked in Example 2.7 by hand. A pipeline with a gas gravity of 0.784 has inlet pipeline conditions of 195°F and 1,050 psia. The gas flowing through the pipeline is cooled by the surrounding water to a temperature of 38°F. The gas also experiences a pressure drop to 950 psia. Gas exits the pipeline at a rate of 3.2 MMscf/D. The pipeline produces condensate at a rate of 25 B/D, with an average density of 300 lbm/bbl and an average molecular weight of 90 lbm/lbm mol. Produced water enters the pipeline at a rate of 0.25 B/D. Determine the rate of MeOH and glycol injection needed to prevent hydrate formation in the pipeline.

Solution. **Fig. 2.12** is a copy of HYDCALC that highlights the data input that is needed to run the program. All required data are provided in the example with the exception of gas gravity. Gas grav-

*Personal communication with P.J. Shuler, Chevron Production Technology Co., La Habra, California (10 June 1997).

a:\excel7\hydcalV7.xls disk 2			P.J. Shuler	CPTC	5/27/97
			CTN 694-7572, PJSH		
HYDCALC Version 2		for Excel 7.0			
INHIBITOR REQUIREMENT CALCULATION			**Inputs**		
FOR A WET GAS FLOWLINE					
USER INPUTS (in red)					
Bottomhole Pressure	1050	psia	.===> starting high pressure		
Cold Line Pressure	950	psia	.===> pressure where hydrates		
Bottomhole Temperature	195	°F	.===> starting high temperature		
Cold Temperature	38	°F	.===> temperature where hydrates		
Gas gravity	0.784				
Gas Rate	3.2	MMscf/D			
Condensate Rate	25	B/D			
Formation Water Rate ??	0.25	B/D H_2O	SUMMARY OF RESULTS		
Calculated Condensed Water	5.7	B/D H_2O			
Total Water to Treat	5.9	B/D H_2O	Methanol Injection Rate	134.9	gal/D
			at		
			Methanol Rate/MMscf	42.2	gal/MMscf
CALCULATION WORKSHEET			(pure MeOH at 77°F)		
Water in hot gas	626.2	lbm/MMscf	MEG Injection Rate	190.0	gal/D
Water in cold gas	6.5	lbm/MMscf	(pure MEG)		
WATER CONDENSED	619.8	lbm/MMscf	MEG Rate/MMscf	59.4	gal/MMscf
Total Water CONDENSED	1983	lbm/D			
in the line	5.7	B/D H_2O			
Total water (from above)	5.9	B/D H_2O			
Hydrate temperature of gas	65.0	°F			
Freeze depression required	27.0	°F			
wt% methanol	27.0	%	**Summary of Results**		
needed in water phase					
wt% MEG	45.6	%			
needed in water phase					
Vapor to liquid	0.9162	lbm/MMscf/			
composition ratio		% in water			
Methanol in gas	24.77	lbm/MMscf			
MEG in gas	0	lbm/MMscf			
Methanol into condensate	37.5	lbm/D			
MEG into condensate	22.5	lbm/D			
Methanol to protect	767	lbm/D			
water phase					
MEG to protect	1,735	lbm/D			
water phase					
TOTALS					
Methanol to protect	767	lbm/D	MEG to protect	1,735	lbm/D
water phase			water phase		
Methanol going to gas	79	lbm/D	MEG in gas	0	lbm/MMscf
Methanol into condensate	37.5	lbm/D	MEG into condensate	22.5	lbm/D
TOTAL Methanol Rate	884	lbm/D	TOTAL MEG Rate	1,758	lbm/D
Methanol Injection Rate	134.9	gal/D	MEG Injection Rate	190.0	gal/D
(pure MeOH at 77F)			(pure MEG)		
Methanol Rate/MMscf	42.2	gal/MMscf		59.4	gal/MMscf

Fig 2.12—Example 2.8 calculated through HYDALC.

TABLE 2.7—COMPARISON OF HAND WITH COMPUTER CALCULATIONS FOR SECOND ESTIMATION METHOD

Calculated Quantity	Hand Method Result With Rules of Thumb	HYDCALC Result
Water condensed, lbm/MMscf	591	619.8
MeOH in water, lbm/MMscf	228.7	239.7
MeOH in gas, lbm/MMscf	27	24.7
MeOH in condensate, lbm/MMscf	11.7	11.7
Total MeOH injection, lbm/MMscf	267.4	276.25
Total MeOH injection, gal/MMscf	40.3	42.2

ity was calculated in Step 1 of Example 2.7 to be 0.784. Fig. 2.12 displays all input data and results.

The amount of MeOH injected is 42.2 gal/MMscf, and the amount of glycol injected is 59.4 gal/MMscf.

For ease of use, the engineer may turn to HYDCALC to perform the second approximation calculation. The following section provides accuracy and limitations of both HYDCALC and the hand-calculation methods, which are vital to their use.

2.6.6 Accuracy, Limitations, and Extensions for Second Estimation Method. **Table 2.7** contains a comparison of the previous results with the hand-calculation method and the HYDCALC method.

While the hand calculation and the computer program provide only slightly different results, both include inaccuracies. For example, while it is possible to obtain more significant figures with HYDCALC than with the charts in the hand method, HYDCALC inaccuracies are those of the charts on which HYDCALC is based.

By use of HYDCALC, it was estimated that 27 wt% MeOH was required in the water phase to inhibit hydrate formation in the pipeline, while measurements by Robinson and Ng[17] show that only 20 wt% MeOH was required for inhibition at the same gas composition, temperature, and pressure of Examples 2.7 and 2.8.

The major inaccuracies in the second estimation method are in the gas-gravity hydrate-formation conditions, which are only accurate to ±7°F or to ±500 psia. In the Hammerschmidt equation, the inhibitor temperature depression ΔT is accurate to ±5%. With such inaccuracies, the amount of MeOH or glycol injection could be in error by 100% or more. The principal virtue of the second estimation method is ease of calculation rather than accuracy.

A second limitation is that the method was generated for gases without H_2S, which represents the case for many gases in the Gulf of Mexico. Baillie and Wichert[18] proposed a modification of the gas-gravity method for sour gases.

2.7 Most Accurate Calculation of Hydrate Formation and Inhibition

If the HYDCALC results indicate that hydrate formation will occur without inhibition, the engineer should elect to do further, more accurate calculations. The most accurate method for hydrate-formation conditions, together with the amount of MeOH needed in the water phase, is available as the final estimation technique in a computer program, HYDOFF. This book provides a user's manual and an example in Appendix B. The method details are too lengthy to include here; the engineer interested in program details is referred to Chap. 5 in Ref. 8.

Sec. 2.7 provides examples for the most accurate methods for the following calculations:

- Calculation of hydrate formation and inhibition in water (Sec. 2.7.1).
- Conversion of MeOH to MEG concentration in water phase (Sec. 2.7.2).
- Calculation of solubility of MeOH and MEG in the gas (Sec. 2.7.3).
- Calculation of solubility of MeOH and MEG in condensate (Sec. 2.7.4).

2.7.1 Hydrate Formation and Inhibitor Amounts in Water Phase. HYDOFF is a PC-compatible computer program provided on the disk included with this book. The program enables the user to determine hydrate-formation conditions and the amount of inhibitor needed in the free-water phase. As a minimum, a 386-IBM computer with 2 megabytes of RAM is required. The program may be executed from either the Windows or the DOS environment.

To use the program, first load both HYDOFF.EXE and FEED.DAT from the accompanying 3.5-in. disk onto a hard drive. Appendix B provides a user's manual with several examples of the use of HYDOFF. Example 2.9 illustrates the simplest (and perhaps the most beneficial) use of HYDOFF.

Example 2.9: Use of HYDOFF To Obtain Hydrate-Formation and -Prevention Conditions. Find the hydrate-formation pressure of the following natural gas at 38°F and the amount of MeOH in the water phase to inhibit hydrates at 38°F and 1,000 psia. The gas composition is methane = 71.60 mol%, ethane = 4.73 mol%, propane = 1.94 mol%, *n*-butane = 0.79 mol%, *n*-pentane = 0.79 mol%, carbon dioxide = 14.19 mol%, and nitrogen = 5.96 mol%.

Solution. The gas in this example has the same composition as the gas in Examples 2.7 and 2.8, so the results provide a comparison with hand and HYDCALC computer calculations of the gas-gravity method (Sec. 2.6.1) and the Hammerschmidt equation (Sec. 2.6.2).

For convenience with multiple calculations, the reader may wish to edit the program FEED.DAT to reflect the gas composition of the problem. Modification of the FEED.DAT program is done in a flat text editor, such as Notepad in Windows, by changing the composition of each component to that of the example gas and saving the result. However, it is not necessary to use FEED.DAT; the gas composition may be entered as part of the HYDOFF program.

In the following solution, each input from the user is underlined.

1. From Windows or in the proper directory, click on, or type HYDOFF; press Enter.
2. After reading the title screen, press Enter.
3. At the "Units" screen, press 1 (to choose °F and psia) then Enter.
4. At the FEED.DAT question screen, press Y and Enter if you wish to use the data in FEED.DAT, or N and Enter if you wish to enter the gas composition in HYDOFF by hand. The remainder of this example is written assuming that the user will enter the gas composition in HYDOFF rather than use FEED.DAT. The use of FEED.DAT is simpler and should be considered for multiple calculations with the same gas.
5. The next screen asks for the number of components present (excluding water). Type 7 and Enter.
6. The next screen requests a list of the gas components present, coded by numbers shown on the screen. Type 1, 2, 3, 5, 7, 8, and 9 (in that order, separating the entries by commas) and then Enter.
7. The next series of screens requests the input of the mole fractions of each component.

Methane 0.7160 Enter.
Ethane 0.0473 Enter.
Propane 0.0194 Enter.
n-Butane 0.0079 Enter.
Nitrogen 0.0596 Enter.
Carbon Dioxide 0.1419 Enter.
n-Pentane 0.0079 Enter.

8. At the "Main" screen, type 1 then Enter.
9. At the "Options" screen, type 1 then Enter.
10. At the screen asking for the required Temperature, type 38 and Enter.
11. Read the hydrate-formation pressure of 230.6 psia (meaning hydrates form at any pressure greater than 230 psia at 38°F for this gas).
12. When asked for another calculation, type N for "No" then Enter.
13. At the "Options" screen type 2, then Enter.
14. At "Inhibitor" screen, type 1, then Enter.
15. At the screen asking for the required temperature, type 38 and Enter.

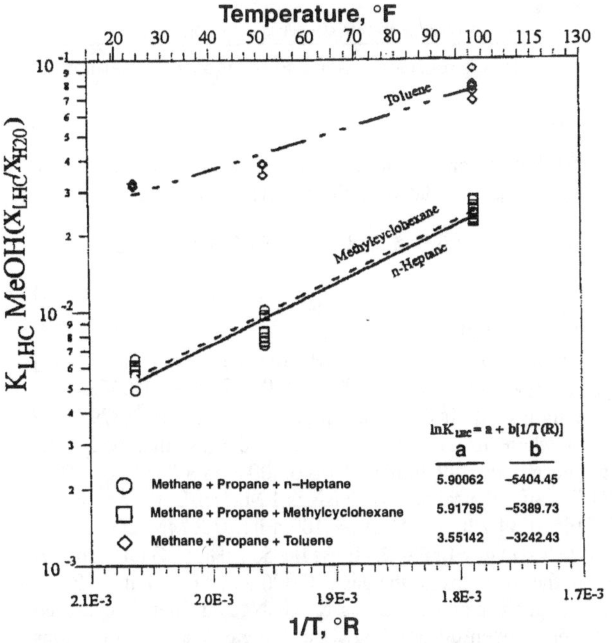

Fig. 2.13—MeOH lost to condensate.[8]

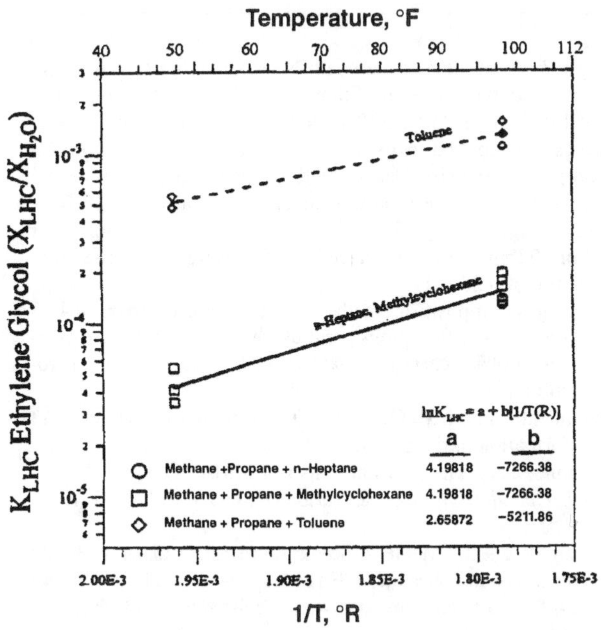

Fig. 2.14—MEG lost to condensate.[8]

16. At the screen to enter the "WEIGHT PERCENT of Methanol," type 22.

17. Read the resulting hydrate condition of 22 wt% MeOH, 38°F, and 972.7 psia.

It may require some trial and error with the use of the program before the correct amount of MeOH is entered that inhibits the system at the temperature and pressure of the example. One starting estimate for the trial-and-error process would be the amount of MeOH predicted by the Hammerschmidt equation (27 wt%) in Example 2.7. Ng and Robinson[19] measured 20 wt% of MeOH in the water required to inhibit hydrates at 38°F and 1,000 psia. A comparison of the measured value with the calculated value (22 wt%) in this example and through the Hammerschmidt equation provides an indication of both the absolute and relative calculation accuracy.

HYDOFF also can be used to predict the uninhibited hydrate-formation temperature at 1,000 psia at 58.5°F, through a similar trial and error process, as compared with 65°F determined by the gas-gravity method. No measurements are available for the uninhibited formation conditions of the gas in this example.

In using HYDOFF, if components heavier than n-decane ($C_{10}H_{22}$) are present, they should be lumped with n-decane because they are all nonhydrate formers.

2.7.2 Replacing MeOH with MEG in the Water Phase. The concentration of inhibiting MEG in the water phase can be determined from MeOH concentration with a simple correlation of inhibitors:

$$\text{wt\% MEG} = -1.209 + 2.34\,(\text{wt\% MeOH}) - 0.052\,(\text{wt\% MeOH})^2 + 0.0008(\text{wt\% MeOH})^3 \quad \text{(2.4)}$$

To use Eq. 2.4, first determine the amount of MeOH required by use of HYDOFF, as in Example 2.9. Insert the amount of MeOH in Eq. 2.4 to determine the amount of MEG needed in water to inhibit hydrates. Eq. 2.4 should be used for the free-water phase only. Example 2.10 provides a summary calculation of all the procedures in Sec. 2.7.

2.7.3 Solubility of MeOH and MEG in the Gas. Fig. 2.10 is a fit of recent measurements by Ng and Chen[20] for K^v_{MeOH}, defined as the ratio of MeOH mole fraction in gas relative to water ($\equiv y_{\text{MeOH}}/x_{\text{MeOH}}$ in H_2O). Once the mole fraction of MeOH in water is determined with HYDOFF, it may be multiplied by K^v_{MeOH} to obtain the mole fraction of MeOH in the gas. As can be determined by Fig. 2.10, the solubility in the water is only slightly affected by pressure over the range from 1,000 to 3,000 psia at offshore temperatures. For a conservative estimate, the 3,000 psia line is recommended:

$$K^v_{\text{MeOH}} = \exp\{5.706 - 5738 \times [1/T(°R)]\}. \quad \text{(2.5)}$$

Fig. 2.11 provides an estimation of MEG dissolved in gas at 1,000 psig from the data of Polderman.[21] As indicated in the figure, the amount of MEG in the vapor is very small; Ng and Chen[20] measure a negligible MEG concentration in the vapor as a comparison. Example 2.10 (Sec. 2.7.5) provides a summary calculation of all the procedures in Sec. 2.7.

2.7.4 Solubility of MeOH and MEG in the Condensate. Fig. 2.13 is a fit of measurements by Ng and Chen[20] for K^L_{MeOH}, defined as the MeOH mole fraction in condensate relative to water ($\equiv x_{\text{MeOH}}$ in HC/x_{MeOH} in H_2O). Once the mole fraction of MeOH in water is determined, it may be multiplied by K^L_{MeOH} to obtain the mole fraction of MeOH in the condensate. In Fig. 2.13 all lines are pressure independent and the toluene line should not apply because of the absence of such compounds in typical condensates. The fit for the solubility of MeOH in condensates of methane, propane, and n-heptane is recommended:

$$K^L_{\text{MeOH}} = \exp\{5.90 - 5404.5 \times [1/T(°R)]\}. \quad \text{(2.6)}$$

Fig. 2.14 shows similar measurements by Ng and Chen[20] that specify the solubility for MEG in the condensate by use of K^L_{MEG} defined as the MEG mole fraction in condensate relative to water ($\equiv x_{\text{MEG in HC}}/x_{\text{MEG in H2O}}$). Note that the K^L_{MEG} values are two orders of magnitude lower than K^L_{MeOH} values. No pressure dependence is observed, and the line for MEG solubility in methane, propane, and n-heptane (or methylcyclohexane) is recommended because toluene is not in condensate:

$$K^L_{\text{MEG}} = \exp\{4.20 - 7.266.4 \times [1/T(°R)]\}. \quad \text{(2.7)}$$

Example 2.10 provides a summary calculation of all the procedures in Sec. 2.7.

2.7.5 Best Calculation Technique for MeOH or MEG Injection. The following example is identical to that of Examples 2.7 and 2.8, with the exception that both MeOH and MEG injection are calculated for comparison of each inhibitor with the less accurate methods of Sec. 2.6.

Example 2.10: Most Accurate Inhibitor Injection Calculation. A pipeline with the gas composition described next has inlet conditions of 195°F and 1,050 psia. The flowing gas is cooled by the surroundings to a temperature of 38°F. The gas also experiences a pressure drop to 950 psia. Gas exits the pipeline at a rate of 3.2 MMscf/D. The pipeline produces condensate at a rate of 25 B/D, with an average density of 300 lbm/bbl and an average molecular weight of 90 lbm/lbmole. Produced salt-free water enters the pipeline at a rate of 0.25 B/D.

Natural gas composition: methane = 71.60 mol%, ethane = 4.73 mol%, propane = 1.94 mol%, *n*-butane = 0.79 mol%, *n*-pentane = 0.79 mol%, carbon dioxide = 14.19 mol%, and nitrogen = 5.96 mol%.

Find the rate of both MeOH and MEG injection needed to prevent hydrate formation in the pipeline.

Solution. Basis: The solution basis is 1 MMscf/D.

Step 1. Calculate the Concentration of MeOH and MEG in the Water Phase.

In Example 2.9 the MeOH concentration was calculated to be 22 wt% of the free-water phase at 38°F and 1,000 psia for this gas. By use of Eq. 2.4, the MEG concentration was calculated at 33.6 wt% in the water phase.

Step 2. Calculate the Mass of Liquid H_2O/MMscf of Natural Gas.

1. Calculate mass of condensed H_2O. Use the water content chart (Fig. 2.9), to calculate the water in the vapor/MMscf. The inlet gas (at 1,050 psia and 195°F) water content is read as 600 lbm/MMscf. The outlet gas (at 950 psia and 38°F) water content is read as 9 lbm/MMscf. The mass of liquid water resulting from condensation is

$$\frac{600\ \text{lbm}}{\text{MMscf}} - \frac{9\ \text{lbm}}{\text{MMscf}} = \frac{591\ \text{lbm}}{\text{MMscf}}.$$

2. Calculate mass of produced H_2O flowing into the line. Convert the produced water of 0.25 B/D to the basis of lbm/MMscf:

$$(0.25\ \text{B/D H}_2\text{O})\left(\frac{42\ \text{gal}}{\text{bbl}}\right)\left(\frac{8.34\ \text{lbm}}{\text{gal}}\right)\left(\frac{1\text{D}}{3.2\ \text{MMscf}}\right)$$

$$= 27.4\ \frac{\text{lbmH}_2\text{O}}{\text{MMscf}}.$$

3. Total mass of water/MMscf gas. Add the condensed and produced water:

$$\frac{591\ \text{lbm}}{\text{MMscf}} + \frac{27.4\ \text{lbm}}{\text{MMscf}} = \frac{618.4\ \text{lbm}}{\text{MMscf}}.$$

Step 3. Calculate the Rate of MeOH and MEG Injection. MeOH and MEG can exist in three phases: water, gas, and condensate. The total masses of MeOH and MEG injected per MMscf are calculated as follows:

1. Calculate amount of MeOH and MEG in the water phase.

a. 22.0 wt% MeOH is required to inhibit the free-water phase, and the mass of water/MMscf was calculated at 618.4 lbm. The mass of MeOH in pounds in the free-water phase per MMscf is

$$22\ \text{wt\%} = \frac{m_{\text{MeOH}}}{m_{\text{MeOH}} + 618\ \text{H}_2\text{O}} \times 100\%.$$

Solving for $m_{\text{MeOH}} = 174.4$ lbm MeOH/MMscf in the water phase.

b. In Step 1, 33.6.0 wt% MEG is required to inhibit the free-water phase, and the mass of water/MMscf was calculated at 618.4 lbm in Step 3. The mass of MEG in pounds in the free-water phase per MMscf is

$$33.6\ \text{wt\%} = \frac{m_{\text{MEG}}}{m_{\text{MEG}} + 618\ \text{H}_2\text{O}} \times 100\%.$$

Solving for $m_{\text{MEG}} = 313.1$ lbm MEG/MMscf in the water phase.

2. Calculate amount of MeOH and MEG lost to the gas.

a. MeOH lost to gas. The mole fraction MeOH in the free-water phase is

$$x_{\text{MeOH}} = \frac{174.4\ \text{lbm MeOH}/(32\ \text{lbm/lbm mol MeOH})}{174.4\ \text{lbm MeOH}/(32\ \text{lbm/lbm mol MeOH}) + 618.4\ \text{lbm H}_2\text{O}/(18\ \text{lbm/lbmmol H}_2\text{O})}$$

$$= 0.137.$$

The distribution constant of MeOH in the gas is calculated at 38°F (497.7°R) by Eq. 2.5, relative to the MeOH in the water

$$K^v_{\text{MeOH}} = \exp\{5.706 - 5,738 \times [1/497.7°\text{R}]\} = 0.00296, \quad (2.5)$$

where °R = °F + 459.69.

The mole fraction of MeOH in the vapor is $y_{\text{MeOH}} = K^v_{\text{MeOH}} \cdot x_{\text{MeOH}}$ or $y_{\text{MeOH}} = 0.00296 \times 0.137 = 0.0004055$.

The daily gas rate is 8,432 lbm mol (= 3.2×10^6 scf/(379.5 scf/lbm mol), where an scf is at 14.7 psia and 60°F), so that the MeOH lost to the gas is 3.42 lbm mol (= 0.0004055 × 8432) or 109.4 lbm/D. Because the calculation basis is 1 MMscf/D, the amount of MeOH lost is 34.2 lbm/MMscf (= 109.4 lbm/3.2 MMscf).

b. MEG lost to gas. In Fig. 2.11, use the 50 wt% MEG line to determine that the MEG lost to the gas is 0.006 lbm/MMscf at 38°F and 1,000 psig; such an amount is negligible. Ng and Chen[20] measured a negligible concentration of MEG in the gas phase at conditions similar to those of this problem.

3. Calculate amount of MeOH and MEG lost to the condensate.

a. MeOH lost to the condensate. The distribution of MeOH in the condensate is calculated with

$$K^L_{\text{MeOH}} = \exp\{5.90 - 5404.5 \times [1/497.7°\text{R}]\} = 0.00702, \quad (2.6)$$

where °R = °F + 459.69.

The mole fraction MeOH in condensate is $z_{\text{MeOH}} = K^L_{\text{MeOH}} \times x_{\text{MeOH}}$ or $z_{\text{MeOH}} = 0.00702 \times 0.137 = 0.0009617$.

The condensate rate is 26.0 lbm mol/MMscf (= 25 B/D × 300 lbm/bbl × 1 lbm mol/90 lbm × 1D/3.2 MMscf); therefore, the amount of MeOH in condensate is 0.025 lbm mol/MMscf (= 0.0009617 × 26/(1 − 0.009617)) or 0.8 lbm/MMscf).

b. MEG lost to condensate. The mole fraction MEG in the water phase is calculated as

$$x_{\text{MEG}} = \frac{313.1\ \text{lbm MEG}/(62\ \text{lbm/lbm mol MEG})}{313.1\ \text{lbm MEG}/(62\ \text{lbm/lbm mol MEG}) + 618.4\ \text{lbm H}_2\text{O}/(18\ \text{lbm/lbm mol H}_2\text{O})}$$

$$= 0.128.$$

The distribution of MEG between the aqueous liquid and condensate is given by

$$K^L_{\text{MEG}} = \exp\{4.20 - 7.266.4 \times [1/497.7°\text{R}]\}$$

$$= 3.04 \times 10^{-5}. \quad (2.7)$$

The mole fraction MEG in condensate is $z_{\text{MEG}} = K^L_{\text{MEG}}\, x_{\text{MEG}}$ calculated as 3.8×10^{-6} (= 3.04×10^{-5} 0.128). The condensate rate is 26.0 lbm mol/MMscf (= 25 B/D × 300 lbm/bbl × 1 lbm mol/90 lbm × 1D/3.2 MMscf) so that the amount of MEG in condensate is 9.9×10^{-5} lbm mol/MMscf [= 0.0000038 × 26/(1 − 0.0000038)] or 0.0061 lbm/MMscf).

4. Calculate the total amount of MeOH/MMscf and MEG/MMscf with **Table 2.8.**

TABLE 2.8—COMPARISON OF MeOH AND MEG REQUIRED

	MeOH	MEG
In water, lbm/MMscf	174.4	313.1
In gas, lbm/MMscf	34.2	0.006
In condensate, lbm/MMscf	0.8	0.0061
Total, lbm/MMscf	209.4	313.11
Total, gal/MMscf	31.5	33.3

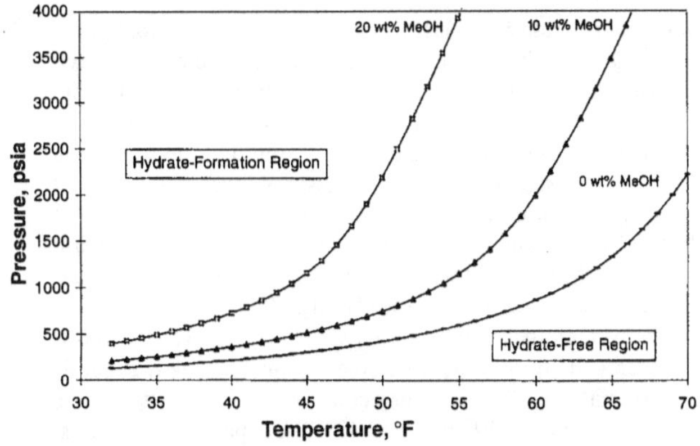

Fig. 2.15—Dog lake field—hydrate curves. (Personal communication with J.L. Todd, 18 June 1997.)

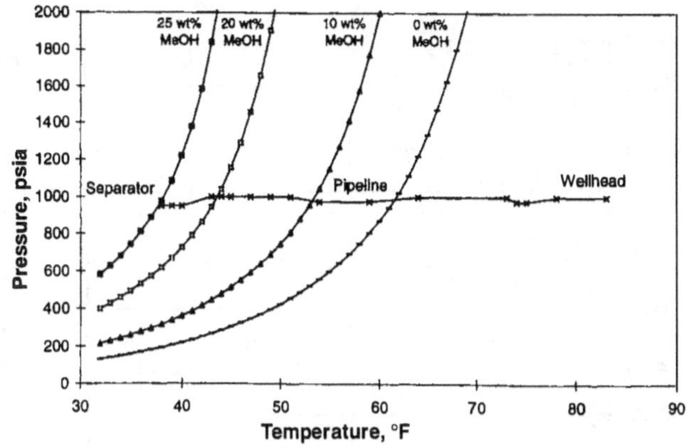

Fig. 2.16—Dog Lake field—original conditions. (Personal communication with J.L. Todd, 18 June 1997).

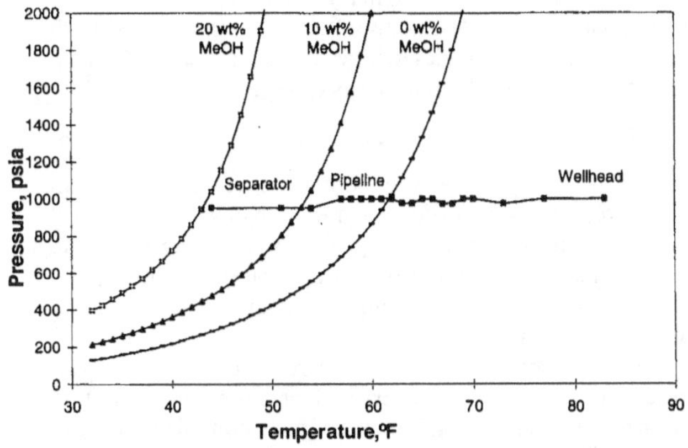

Fig. 2.17—Dog Lake field with burial. (Personal communication with J.L. Todd, 18 June 1997).

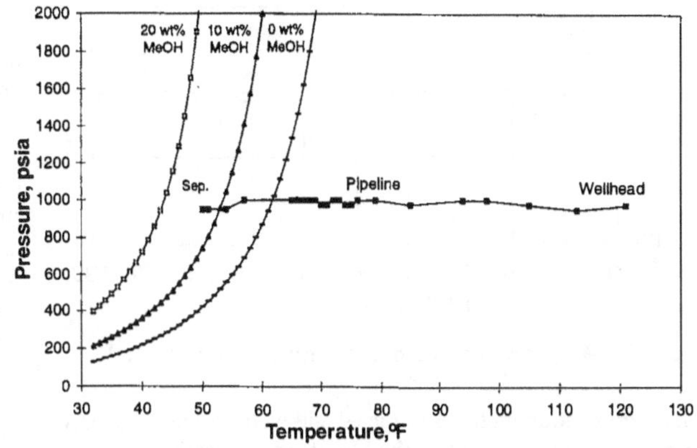

Fig. 2.18—Dog Lake field with burial and heating. (Sep.=sepera-tor) (Personal communication with J.L. Todd, 18 June 1997.)

The example illustrates that, for this gas condition, the injection amounts of MeOH and MEG are comparable. The more precise calculation shown here, however, represents a considerable savings in the amount of MeOH injected (31.5 vs. 42.2 gal/MMscf in the second estimation method shown in Example 2.8).

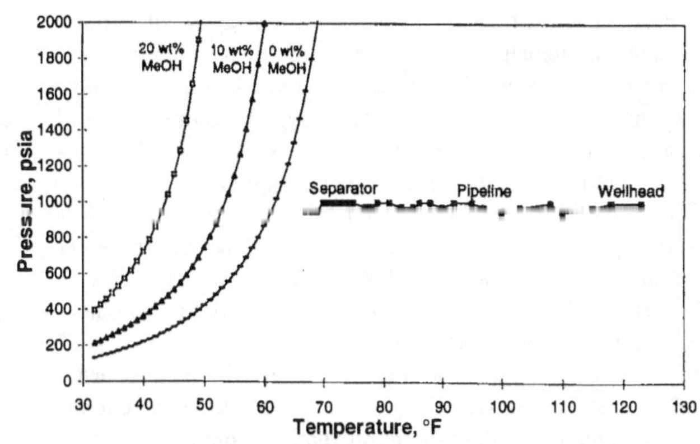

Fig. 2.19—Dog Lake field with burial, heating, and insulation. (Personal communication with J.L. Todd, 18 June 1997.)

Case Study 2.1: Prevention of Hydrates in Dog Lake Field Pipeline. As a summary of the thermodynamic hydrate-prevention methods, consider the steps taken to prohibit hydrates in the Dog Lake field export pipeline in Louisiana.[7] During the winter months, hydrates formed in the line, which traverses land and shallow water (a marsh).

Hydrate-formation conditions, shown in **Fig. 2.15,** are calculated by HYDOFF with 0, 10, and 20 wt% MeOH in the water phase. The Dog Lake gas composition is 92.1 mol% methane, 3.68 mol% ethane, 1.732 mol% propane, 0.452 mol% *i*-butane, 0.452 mol% *n*-butane, 0.177 mol% *i*-pentane, 0.114 mol% *n*-pentane, 0.112 mol% hexane, 0.051 mol% heptane, 0.029 mol% octane, 0.517 mol% nitrogen, and 0.574 mol% carbon dioxide.

The pipeline pressure and temperature, calculated with PIPEPHASE, were superimposed on the hydrate-formation curve shown in **Fig. 2.16.** Gas leaves the wellhead at 1,000 psia and 85°F, far from hydrate-forming conditions. As the gas moves down the pipeline, it begins to cool toward ambient temperatures. Once the temperature reaches approximately 63°F, hydrates form, so MeOH must be added. The figure shows pipeline conditions and the hydrate-formation curves for various concentrations of MeOH, indicating that 25 wt% MeOH in water is needed to inhibit hydrates.

Todd *et al.*[7] indicate that despite large quantities of MeOH injected for hydrate prevention, 110 hydrate incidents occurred in the line during winter of 1995–96 at a cost of U.S. $325,000. Combinations of four alternative hydrate-prevention methods were considered: (1) burying the pipeline, (2) heating the gas at the wellhead, (3) insulating the pipeline, and (4) adding MeOH. The details of each prevention measure are considered next.

1. Burying the pipeline. Portions of the Dog Lake pipeline were built over a stretch of marsh. The exposure to winter ambient temperatures caused rapid reductions in the gas temperature. Burying the pipeline would protect it from low environmental temperatures because of the higher earth temperatures. **Fig. 2.17** displays the temperature increase in the pipeline after exposed areas were buried relative to the exposed pipeline in Fig. 2.16. With pipeline burial, the need for MeOH in the water phase was reduced from 25 to less than 19 wt%.

2. Wellhead heat addition. Catalytic in-line heaters could be installed at the wellhead to increase the gas temperature to 125°F. **Fig. 2.18** shows the pipeline temperature increase caused by the

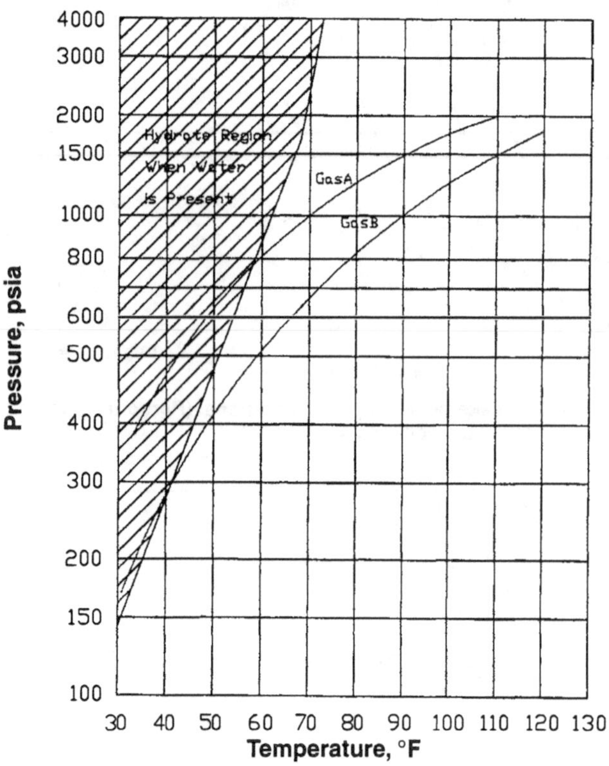

Fig. 2.20—Expansions into hydrate-formation region.[11]

combined prevention methods of burial and wellhead heating. Use of these two methods permitted the MeOH concentration to be reduced to approximately 14 wt% to prevent hydrate formation in the line. It should be noted that heating may increase the amount of corrosion in the line.

3. Insulation. Insulation of exposed areas near the wellhead and battery maintains higher pipeline temperatures, thereby reducing the amount of MeOH needed for hydrate inhibition. **Fig. 2.19** displays the temperature increase in the buried and heated pipeline when exposed pipes were insulated. The pipeline is now outside the hydrate-formation region, and MeOH addition is no longer needed.

4. MeOH addition. Continued MeOH injection could be done at an cost of approximately U.S. $1.50 to $2.00/gal. The cost of MeOH to an offshore platform was U.S. $2.00/gal during the 1996–97 winter. Because MeOH recovery may not be economical, MeOH is normally considered an operating cost.

This case study illustrates how combinations of pipeline burial, insulation, heating, and MeOH injection can be used to prevent hydrates. The selection of the hydrate-prevention scheme(s) is then a matter of competing capital and operating costs, as considered in Chap. 4 on economics.

2.8 Hydrate Formation During Expansion Through Valves or Restrictions

When water-wet gas expands rapidly through a valve, orifice, or other restriction, hydrates form because of rapid gas cooling caused by Joule-Thomson expansion. Hydrate formation with rapid expansion from a wet line commonly occurs in fuel gas or instrument gas lines, as Example 2.13 indicates. Hydrate formation with high pressure drops can occur in well testing, startup, and gas-lift operations, even when the initial temperature is high, if the pressure drop is very large. This section provides methods to determine when hydrates will form on rapid expansion. A rough estimation method (Sec. 2.8.1) is followed by a more accurate but resource intensive method (Sec. 2.8.2), concluding with prevention techniques in Sec. 2.8.3.

Figs. 2.4 and 2.5 are schematics of the pressure and temperature of a pipeline production stream during normal flow with entry into the hydrate-formation region. If the gas expands more rapidly, the normal pipeline cooling curve of Figs. 2.4 and 2.5 takes on a much steeper slope, but the hydrate-formation line remains the same. **Fig. 2.20** shows two rapid expansion curves for a 0.6 gravity gas. Intersections of the gas-expansion curves with the hydrate-formation line limits the expansion discharge pressures from two different high initial pressure/temperature conditions.

In Fig. 2.20, the curves determine the restriction downstream pressure at which hydrate blockages form for a given upstream pressure and temperature. Gas A expands from 2,000 psia and 110°F until it strikes the hydrate-formation curve at 780 psia and 57°F, so 780 psia represents the limit to hydrate-free expansion. Gas B expands from 1,800 psia and 120°F to intersect the hydrate-formation curve at a limiting pressure of 290 psia and 42°F. In expansion processes, the upstream temperature and pressure are known, the discharge temperature is almost never known, and the discharge pressure is normally set by a downstream vessel or pressure drop.

Cooling curves, such as the two in Fig. 2.20, were determined for constant enthalpy (or Joule-Thomson) expansions, obtained from the first law of thermodynamics for a system flowing at steady-state, neglecting kinetic and potential energy changes:

$$\Delta H = Q - W_s \, , \qquad (2.8)$$

where ΔH = enthalpy difference across the restriction (downstream-upstream), Q = heat added, and W_s = shaft work obtained at the restriction. Offshore restrictions (e.g., valves or orifices) have no shaft work and, because rapid flow approximates adiabatic operation, both W_s and Q are zero, resulting in constant enthalpy ($\Delta H = 0$) operation on expansion.

Because of the constant-enthalpy requirement, rapid gas expansion results in cooling except at very high pressures, where heating occurs on expansion resulting from a compressibility decrease with temperature. The upstream pressure at which the system changes from heating to cooling on expansion is called the Joule-Thomson inversion pressure.

Rule-of-Thumb 7. Natural gases cool on expansion from pressures less than 6,000 psia; at more than 6,000 psia, the temperature increases on expansion. Virtually all offshore gas processes cool on expansion because only a few reservoirs and no current pipelines or process conditions are at pressures greater than 6,000 psia.

G.G. Brown,[22] who constructed the first natural gas enthalpy/entropy charts, determined Rule-of-Thumb 7.

2.8.1 Rapid Calculation of Hydrate-Free Expansion Limits.

Katz[11] generated charts to determine the hydrate-free limit to gas expansion, combining the gas-gravity chart (Fig. 2.8) to obtain the hydrate-formation line in Fig. 2.20, with gas enthalpy-entropy charts by Brown[22] to obtain the cooling line.

Cautioning that the charts applied to gases of limited compositions, Katz provided expansion charts for gases of 0.6, 0.7, and 0.8 gravities, shown in **Figs. 2.21 through 2.23,** respectively. The abscissa (or x axis) in each figure represents the lowest downstream pressure without hydrate formation, given the upstream pressure on the ordinate (y axis) and the upstream temperature (a parameter on each line).

It should be noted that maxima in Figs. 2.21 through 2.23 occur at the upstream pressure of 6,000 psia, the Joule-Thomson inversion pressure. This provides a further validation of Rule-of-Thumb 7.

The following three examples for chart use are from Katz's[11] original work.

Example 2.11a: Maximum Pressure of Gas Expansion. To what pressure may a 0.6-gravity gas at 2,000 psia and 100°F be expanded without danger of hydrate formation?

Solution. From Fig. 2.21, read 1,050 psia.

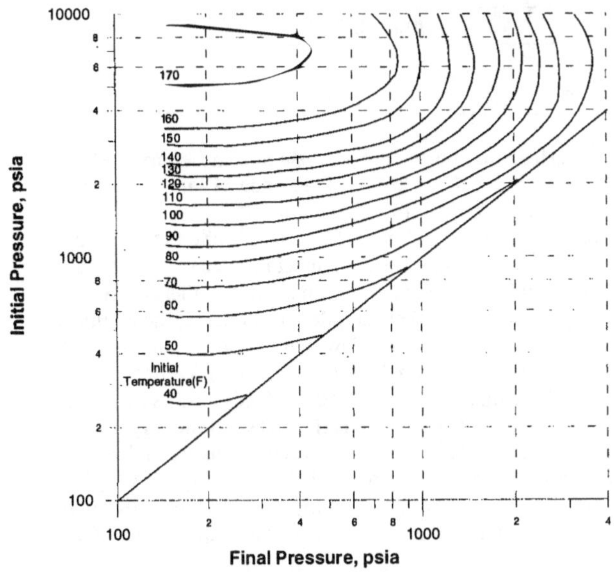

Fig. 2.21—Gas expansion of 0.6-gas-gravity natural gas.[10]

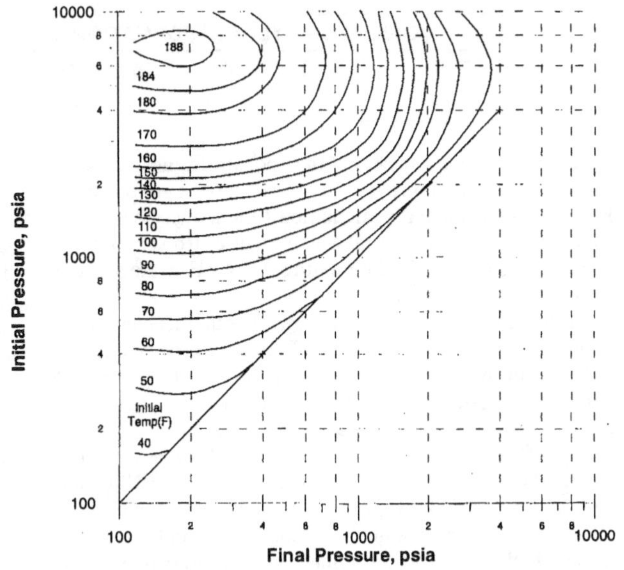

Fig. 2.22—Gas expansion of 0.7-gas-gravity natural gas.[10]

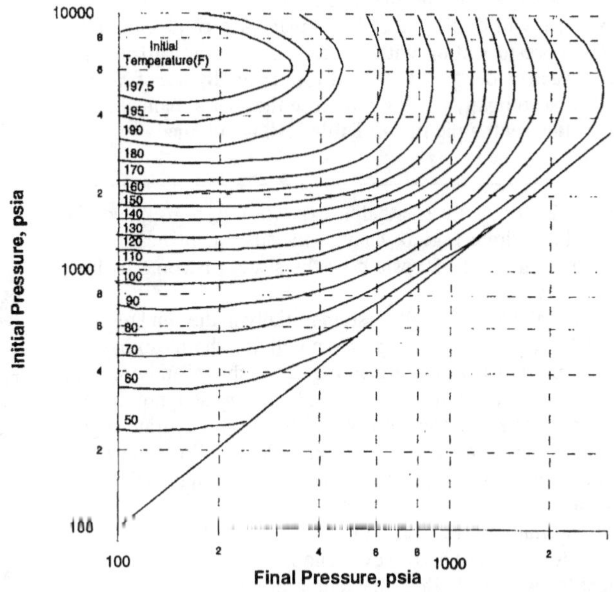

Fig. 2.23—Gas expansion of 0.8-gas-gravity natural gas.[10]

Example 2.11b: Unlimited Gas Expansion. How far may a 0.6-gravity gas at 2,000 psig and 140°F be expanded without hydrate formation?

Solution. Fig. 2.21 shows that there is no intersection with the 140°F isotherm. Hydrates will not form on expansion to atmospheric pressure.

Example 2.11c: Minimum Initial Temperature Before Expansion. A 0.6-gravity gas is to be expanded from 1,500 to 500 psia. What is the minimum initial temperature that will permit the expansion without danger of hydrates?

Solution. From Fig. 2.21, the answer is read as 99°F or greater.

Figs. 2.21 through 2.23 for gas expansion incorporate the inaccuracies of gas-gravity charts from which they were derived. As indicated in Sec. 2.6, the 0.6–gravity chart (used for both hydrate formation and gas expansion) may have inaccuracies of $\pm$ 500 psia. Loh et al.[23] tested accuracy limits to these expansion curves and found, for example, that the allowable 0.6-gravity gas expansion from 150°F and 3,500 psia should be 410 psia rather than the value of 700 psia given by Fig. 2.21.

2.8.2 More Accurate Calculation of Hydrate-Free Limits to Gas Expansion. A more accurate computer method is available that uses the same principles indicated in Fig. 2.20. Just as before, for an initial temperature, pressure, and gas composition, the intersection of an isenthalpic ($\Delta H=0$) cooling curve with the hydrate three-phase locus may be determined. In the new method, the isenthalpic line is determined with a modern equation-of-state (EOS), and the program HYDOFF replaces the gas-gravity chart to predict hydrate-formation conditions. While this method requires more resources (namely time and a PC-compatible computer) than the Joule-Thomson charts, it results in higher accuracy and provides an estimation of the amount of MeOH inhibitor required.

To use the more accurate method, first generate the hydrate stability pressure/temperature line as in Fig. 2.20 by use of HYDOFF, as indicated in Sec. 2.7.1. Later, the amount of MeOH injected to displace the hydrate-formation curve to the left can be calculated, as Example 2.13 illustrates.

The computer program XPAND is included with this book for calculation of the Joule-Thomson expansion line that intersects the hydrate-formation line. Appendix B gives a user's manual for XPAND and an example. The expansion line is calculated in Appendix A of Ref. 8 with an EOS by use of the method detailed. Given an inlet temperature, pressure, and gas composition, the program calculates the enthalpy change, ΔH, for a specified outlet pressure and a temperature guess. At incrementally lower pressures, the user changes the outlet temperature guess until a value of $\Delta H=0$ is obtained. The expansion curve results from a plot of the discharge temperature at each lower pressure. For a starting inlet temperature and pressure, a series of discharge points form a curve that intersects the hydrate-formation curve at the limiting temperature and pressure of expansion. Results may be compared with Joule-Thomson results of Figs. 2.21 through 2.23.

Example 2.12: Hydrate Formation on Expansion of a Natural Gas. A simple natural gas consists of 90 mol% CH_4, 7 mol% C_2H_6, and 3 mol% C_3H_8 with free water in a pipeline. Two initial inlet process conditions are considered for expansion across a valve: (1) 68°F and 2,180 psia or (2) 77°F and 2,180 psia. For either condition, is hydrate formation a possibility? Are there process limitations on the expansion from either initial condition to 1,450 psia?

Solution. Before doing any hydrate calculations, one should confirm that this gas is not close to the hydrocarbon dewpoint to elimi-

TABLE 2.9—HYDRATE FORMATION CONDITIONS FOR GAS IN EXAMPLE 2.11

T,°F	32	35	40	45	50	55	60	65	68	71	77
p, psia	119	149	213	303	368	551	718	1,117	1,624	2,509	4,046

nate the possibility of encountering both vapor and liquid hydrocarbon phases. The expansion program was written for a gas phase. A vapor/liquid equilibrium flash calculation (not shown here) indicated that the highest temperature at which a hydrocarbon liquid can occur (the cricondentherm) for this mixture is −44°F, so the process will not form hydrocarbon liquid.

Fig. 2.24 shows the expansion conditions of both inlet conditions for the gas. The remainder of this example concerns the generation of Fig. 2.24 and the processing implications. First, the pressures and temperatures of hydrate formation are calculated with the program HYDOFF as in Sec. 2.7 and **Table 2.9.**

A semilogarithmic interpolation of the Table 2.9 values gives the hydrate-formation point at 70°F when the pressure is 2,180 psia. Therefore, as Fig. 2.24 shows, the initial condition of 68°F and 2,180 psia is within the hydrate-formation region, but the initial conditions of 77°F and 2,180 psia remain in the fluid (vapor/liquid water) region.

If the system at 68°F and 2,180 psia has formed hydrates, consider two means of depressurization. If the system pressure is lowered to 1,450 psia slowly and isothermally ($\Delta T=0$ with substantial heat input), hydrates will dissociate at 1,537 psia. A second, isenthalpic ($\Delta H=0$) depressurization without heating from the surroundings results in much colder gas (39.8°F) at 1,450 psia. By use of XPAND on the disk accompanying this book (see the user's manual in Appendix B) the isenthalpic line in **Table 2.10** is obtained.

As Fig. 2.24 shows, the isenthalpic expansion system extends further into the hydrate region. Only with subsequent heating at a constant pressure of 1,450 psia does the system become hydrate-free at 66.6°F.

A similar calculation for the system initially in the fluid region at 77°F and 2,180 psia shows the problem with isenthalpic expansion. The result, plotted as Line ABC in Fig. 2.24, shows an isenthalpic intersection with the hydrate-formation boundary at approximately 70.5°F, 1,990 psia. To prevent expansion into the hydrate region, four options may be considered, as Example 2.13 illustrates.

1. Limit the final expansion pressure to a value higher than 1,990 psia.
2. Add inhibitor at the restriction inlet.
3. Dehydrate the gas before expansion.
4. Heat the gas to a higher inlet temperature.

Pipeline hydrate plugs are frequently porous so that depressurization from one (downstream) side can result in Joule-Thomson cooling as gas flows through the plug. Expansion across a hydrate plug yields identical results to expansion across a valve. In the initial part of the previous example, rapid isenthalpic expansion from a condition that has a hydrate plug (e.g., 68°F and 2,180 psia) only causes the downstream portion of the plug to progress further into the hydrate region. Heat must be put into the system from the surroundings to dissociate hydrates. Case Studies C.15 and C.17 in Appendix C give field tests that confirm this discussion.

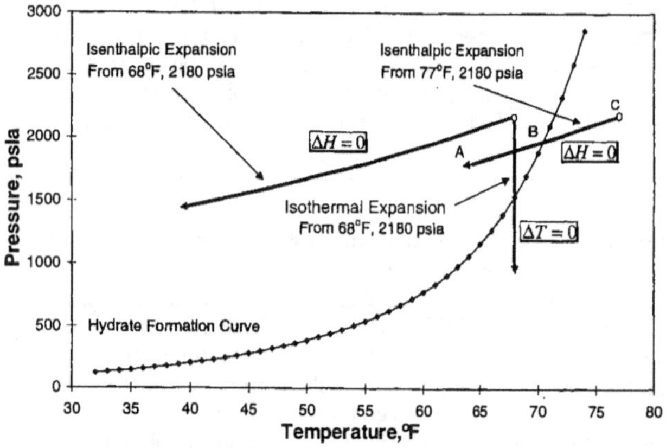

Fig. 2.24—Joule–Thomson cooling through gas expansion.[8]

XPAND has several limitations. First, it is limited to the vapor phase and does not account for the temperature change on expansion of any liquid. If there is a question about whether the system might contain a liquid at either the inlet or discharge, the engineer should calculate the hydrocarbon dewpoint and an isenthalpic flash should be performed to obtain the cooling curve by use of a process simulator package like HYSIM, ASPEN, or PROCESS. Second, because XPAND was generated for only the first six common paraffins (methane, ethane, propane, normal butane, *i*-butane, and *n*-pentane), it cannot be used with nitrogen, acid gases (H_2S or CO_2), or with significant amounts of heavy components. The engineer may group components larger than pentanes into the "pentane plus" fraction of the gas.

2.8.3 Methods To Prevent Hydrate Formation on Expansion. Frequently, gas expansion causes hydrate formation in fuel-gas lines and in instrument gas lines on a platform, which may result in other, larger hydrate problems. In some cases, hydrate formation in a platform instrument gas line has caused system shutdown. Subsequent cooling of the nonflowing pipeline into the hydrate-formation region resulted in a pipeline blockage on resumption of flow.

The following example provides a Sec. 2.8 summary of hydrate prevention during gas expansion, with four methods for hydrate prevention in a fuel-gas line used to supply power to platform compressors.

Example 2.13a: Hydrate Formation During Gas Expansion. An offshore platform design required fuel gas at 300 psia and a rate of 0.02 MMscf/D from a high-pressure flowline at 1,500 psia and 100°F. The inlet flowline was off gas from the first-stage separator (see Fig. 2.6, Example 2.3) so the gas was saturated with water. A control valve was placed on the fuel-gas line from the inlet flowline to provide the required pressure and flow of fuel gas. The mole fraction composition of the components was 0.927 methane, 0.053 ethane, 0.014 propane, 0.0018 *i*-butane, 0.0034 *n*-butane, and 0.0014 *i*-pentane.

Is there a chance that hydrate formation might occur in the fuel gas line? If so, which of the following ways could be used to prevent hydrates?

1. Two-stage expansion with intermediate heat addition.
2. MeOH injection upstream of expansion.
3. Parallel expansions.
4. Drying the inlet gas.

Solution. Examples 2.13b through 2.13g provide the example solution.

Example 2.13b: Hydrate Prediction Through Joule-Thomson Diagrams. By use of the Katz Joule-Thomson expansion diagrams (Figs. 2.21 through 2.23), the minimum initial temperature required for hydrate-free operation can be estimated. This gas is identical to that in Example 2.5, whose gravity is calculated as 0.603. Fig. 2.21 provides an estimate that a 0.6-gravity natural gas must have an initial temperature of 104°F to prevent hydrate formation during gas expansion from 1,500 to 300 psia. Under the current design, the initial temperature of 100°F will cause hydrates to form just downstream of the control valve on fuel-gas takeoff line.

TABLE 2.10—ISENTHALPIC EXPANSION FROM LOW TEMPERATURES IN EXAMPLE 2.11

p, psia	2,100	2,000	1,800	1,600	1,450
T, °F for $\Delta H=0$	68.0	62.4	55.2	46.9	39.8

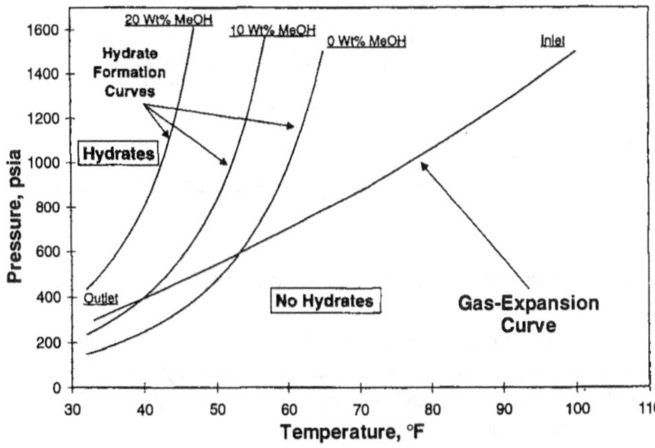

Fig. 2.25—Hydrate-formation curve for single valve expansion.

Example 2.13c: Hydrate Prediction With XPAND and HYDOFF. XPAND was used to calculate the discharge temperature of the natural gas on expansion by use of upstream valve pressure, temperature, and gas composition inputs to calculate the downstream gas temperature at a given discharge pressure. Appendix B is a stepwise XPAND user's manual for the calculation in this example.

Once the expansion pressure/temperature values are obtained, they determine the intersection with hydrate-formation curves (in cluding inhibited curves) generated by HYDOFF. **Fig. 2.25** shows such an intersection for this example.

Note that the expansion line is curved in Fig. 2.25, requiring calculation of several temperatures and pressures along the expansion line. The expansion enters the uninhibited formation region at 53°F, and the final temperature after expansion is calculated as 33°F. For a comparison with 104°F inlet temperature requirement by the Katz Joule-Thomson charts, the inlet temperature with XPAND is calculated to be 108°F for hydrate-free expansion from 1,500 to 300 psia.

Example 2.13d: Hydrate Prevention. After establishing that hydrates form on gas expansion, the platform design was modified to inhibit hydrate formation. Four hydrate-prevention methods were considered: (1) two-stage expansion with heat addition, (2) MeOH addition, (3) parallel expansion, and (4) drying the gas. This section provides details of each prevention method.

Two-Stage Expansion With Heat Addition. In-line heaters could be installed to raise the temperature of the gas outside the hydrate-formation region. In the case considered here, two control valves are used with an in-line heater between them (**Fig. 2.26**). The cooler gas present after the first pressure drop requires heat transfer before flowing through the second valve. The following calculations estimate system pressures and temperatures shown in Fig. 2.26.

In our example, the pressure ratio (p_{in}/p_{out}) is set arbitrarily approximately equal across each control valve, providing 675 psia as the intermediate pressures after the first control valve. By use of XPAND, the temperature of the gas at 675 psia is predicted to be 58°F at the first valve discharge. **Fig. 2.27** shows the gas-expansion conditions and the HYDOFF hydrate-formation curves, demonstrating that the gas is outside the hydrate-formation region after the first pressure drop (Line 1).

In Fig. 2.27, heat is added to the system (Line 2) to raise the temperature to prevent hydrates on gas expansion across the second control valve (Line 3). The heat duty in the exchanger was defined by the temperature increase ($T_3 - T_2$). XPAND was used to estimate a value of T_3 at the second valve inlet, which provided a discharge value T_4 outside the hydrate-formation region. For this example, a T_3 of 68°F is required to maintain the final temperature at 44°F, just above the hydrate-formation region at the second valve discharge pressure (300 psia).

Inspection of Fig. 2.27 shows that initial heating before expansion through a single valve may provide a more economical method to prevent hydrates on expansion. A single control valve and heater would save the capital cost of an additional control valve and may be a better alternative to prevent hydrates on expansion.

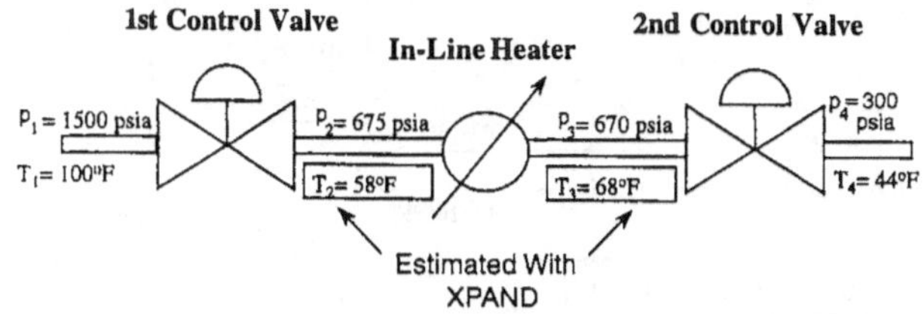

Fig. 2.26—Two-stage gas expansion with heating.

MeOH Addition. MeOH can be injected into the fuel-supply line upstream of the control valve to prevent hydrate formation downstream of the valve. Fig. 2.25 shows that more than 10 wt% MeOH is needed in the free-water phase to prevent hydrate formation. A better estimate of 12 wt% MeOH in the water phase was obtained through interpolation with XPAND and HYDOFF. The total amount of MeOH required for upstream gas injection is calculated with methods of Sec. 2.7, as shown in the next calculations.

Water Condensation With Expansion. Gas flows into the fuel line at a rate of 0.02 MMscf/D. Because the gas is saturated with water, one can calculate the mass of pipeline free water resulting from water dewpoint condensation from Fig. 2.9 (45 lbm H_2O/MMscf in the vapor at 1,500 psia and 100°F and 16 lbm H_2O/MMscf in the vapor at 350 psia and 33°F). The amount of free water that forms from the vapor is 45 lbm/MMscf − 16 lbm/MMscf = 29 lbm/MMscf.

Consequently, the total amount of water, m_W, condensed per day is

$$\frac{29 \text{ lbm H}_2\text{O}}{\text{MMscf}} \times \frac{0.02 \text{ MMscf}}{\text{D}} = 0.58 \text{ lbm/D H}_2\text{O}.$$

Mass of MeOH Required in the Water Phase. The mass of MeOH can be found with the definition of weight percent:

$$W_{\text{MeOH}} = \frac{m_{\text{MeOH}}}{m_{\text{MeOH}} + m_W} \times 100\%,$$

where m_W is weight of water in pounds (0.58 lbm/D).

Solving for 12 wt%, the amount of MeOH, m_{MeOH} = 0.079 lbm MeOH/D.

Mass of MeOH Lost to Condensate and Vapor. The mole fraction of MeOH in the water is found by the equation:

$$x_{\text{MeOH}} = \frac{0.079 \text{ lbm MeOH}/(32 \text{ lbm/lbm mol MeOH})}{0.079 \text{ lbm MeOH}/(32 \text{ lbm/lbm mol MeOH}) + 0.58 \text{ lbm H}_2\text{O}/(18 \text{ lbm/lbm mol H}_2\text{O})}$$

$$= 0.071.$$

The distribution coefficient of MeOH in the gas is calculated at 33.03°F (492.7°R) by Eq. 2.5, relative to the MeOH in the water,

$$K^v_{\text{MeOH}} = \exp\{5.706 - 5738 \times [1/492.7°\text{R}]\} = 0.00263. \quad (2.9)$$

The mole fraction of MeOH in the vapor is $y_{\text{MeOH}} = K^v_{\text{MeOH}}\, x_{\text{MeOH}}$ or $y_{\text{MeOH}} = 0.00263 \times 0.071 = 0.000187$.

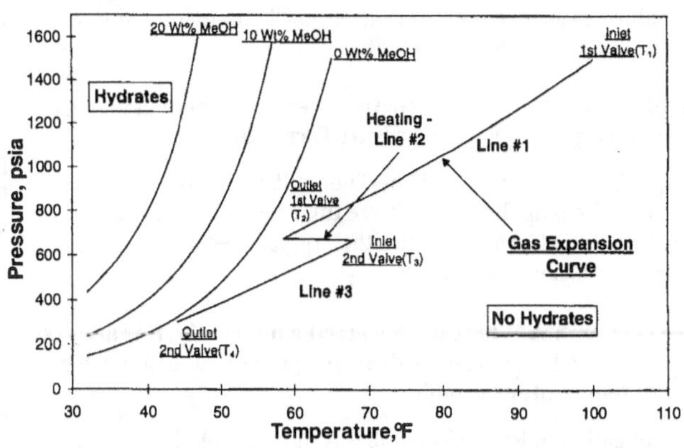

Fig. 2.27—Hydrate curve of two-stage gas expansion with heating.

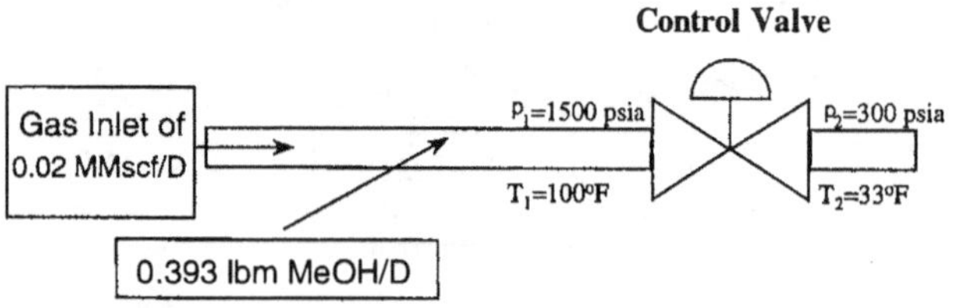

Fig. 2.28—Single-valve gas expansion with heating.

The daily gas rate is 52.7 lbm mol [$= 2.0 \times 10^4$ scf/(379.5 scf/lbm mol)], so that the MeOH lost to the gas is 0.0098 lbm mol/D ($= 0.000187 \times 52.7$) or 0.314 lbm/D.

No condensate is formed in the pipeline, thus no MeOH is lost to the liquid-hydrocarbon phase.

Total Mass of MeOH Needed. The total amount of MeOH injected is the sum of the methanol in the vapor (0.314 lbm), condensate (0), and water (0.079 lbm) = 0.393 lbm MeOH/D (0.06 gal/D) injection required to inhibit the fuel gas line, with injection before the control valve as shown in **Fig. 2.28.**

Example 2.13e: Parallel Gas Expansion. Operating personnel sometimes suggest that fuel-gas lines be placed in parallel to provide more than one gas expansion (**Fig. 2.29**). If one control valve becomes plugged with hydrates and shuts down, the second gas line is then opened while the first line is depressurized for hydrate dissociation. With this technique, flow may be maintained in one fuel-gas line without the need for hydrate inhibition.

Conditions of hydrate formation upon parallel gas expansion are exactly the same as in Fig. 2.25. The capital cost is doubled, however, and there is the risk that the parallel valve may become hydrated before the plug is removed from the initial line. This solution technique addresses the effect of hydrate formation rather than its cause and should be considered a less-than-optimal operating practice.

Example 2.13f: Drying the Inlet Gas. If the gas inlet is dry, hydrates cannot form because of insufficient water. It is good design practice to place both fuel-gas and instrument-gas lines downstream of a triethylene glycol (TEG) drying unit or a molecular sieve adsorption tower. The design of a drying unit is outside the scope of this book but is readily available in standard texts on gas processing (e.g., Ref. 6).

Of the previous four design methods to prevent hydrates in fuel-gas lines, the most satisfactory according to expense and good operating practice is to provide dry inlet gas with a fuel-gas takeoff downstream of the TEG dryer. As Deaton and Frost[24] state in their classic study of hydrate formation and prevention: "The only method found to be completely satisfactory in preventing the formation of hydrates in gas transmission lines is to dehydrate the gas entering the line to a dewpoint low enough to preclude formation of hydrates at any point in the system."

The above study of gas expansion without hydrate formation suggests two additional rules of thumb.

Rule-of-Thumb 8. It is always better to expand a dehydrated gas than a moist gas to prevent hydrate formation.

Example 2.13f illustrated Rule-of-Thumb 8, which typifies instrument- or fuel-gas applications. To use this rule of thumb, it is necessary to be able to dry the gas with either a glycol dehydrator or a molecular-sieve adsorption process.

Rule-of-Thumb 9. Where drying is not a possibility, it is always better to take a large pressure drop at a process condition where the inlet temperature is high.

One application of Rule-of-Thumb 9 is the bottomhole choke, discussed in Ref. 7. A bottomhole choke is a device with a restricted opening that is placed in the lower end of the tubing string to cause a large pressure drop to be taken deep in the wellbore. The warm downhole reservoir heats the gas before it expands, thus preventing hydrates from forming across the expansion. The majority of bottomhole chokes are installed in high-pressure gas wells that produce a low amount of liquids.

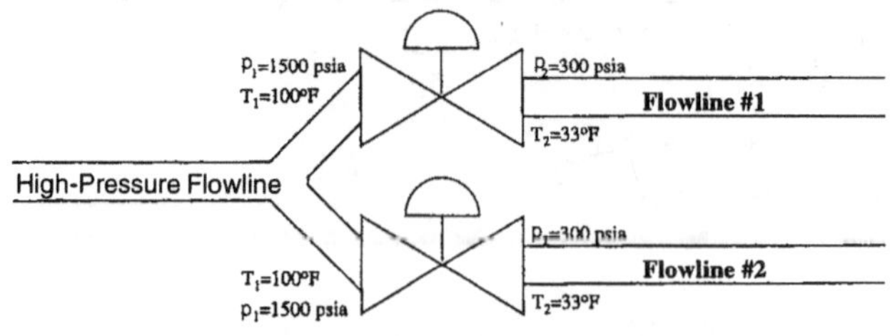

Fig. 2.29—Parallel gas expansion.

2.9 Hydrate Control Through Chemical Inhibition and Heat Management

There are four classic approaches to hydrate inhibition, discussed at the beginning of this chapter: (1) remove water from the system, (2) increase the temperature, (3) decrease the pressure, or (4) insert an antifreeze component, such as an alcohol or glycol. Two additional, new inhibition techniques have been commercialized and are gaining industrial acceptance: (5) the kinetic inhibition method of preventing sizable crystal growth for a period exceeding the free-water residence time in a pipeline and (6) the antiagglomerant method, which uses a surfactant to stabilize the water/hydrate phase as small, emulsified droplets within a liquid hydrocarbon.

Thermodynamic inhibition (Methods 1 through 4) prohibits hydrate formation altogether, while with the newer methods (5 and 6), the system is allowed to exist within the hydrate-stability zone but small hydrate crystals are stabilized for some time period without growing to larger masses. While thermodynamic inhibitors are the standard practice, there are successful commercial instances of kinetic control. The incentive for newer kinetic control methods is a substantial offshore capital-cost reduction by elimination of the need for offshore platform equipment, an operating-cost reduction, and elimination of some environmental concerns. In the future, innovative methods of heat management through heating and insulation may provide thermodynamic protection against hydrates.

Sec. 2.9.1 discusses design and operation with thermodynamic inhibition chemicals (MeOH and MEG). Sec. 2.9.2 discusses design and operation with kinetic inhibitors. Sec. 2.9.3 summarizes the chemical-inhibitor-use guidelines. Sec. 2.9.4 summarizes methods of heat management to retain a high inlet temperature in the fluid region without hydrate formation.

2.9.1 Inhibition With MeOH or MEG. ***MeOH.*** Of all hydrate inhibitors, MeOH is the most widely used. MeOH is also the best and most cost-effective of the alcohols. Hydrate inhibition abilities are less for larger alcohols (i.e., MeOH > ethanol > isopropanol). Typically MeOH is vaporized into the pipeline gas stream so that MeOH flows with the gas and dissolves in any free-water accumulation(s) to prevent hydrate formation.

For MeOH injection in wells, a commercial program, such as WELLTEMP® or WEST®, can be used to predict the flowing-well temperature and pressure (in an identical manner to that used with PIPEPHASE® or OLGA® in Example 2.2). The downhole MeOH injection point is placed at the well depth where the well temperature and pressure are predicted to cross into the hydrate-formation region, for well production, testing, and shut-in conditions. Protection during well testing and shut-in is essential because the flowing-well conditions are normally warm enough to prevent hydrate formation. See Sec. 2.11 for a discussion of hydrates in drilling.

The MeOH amount needed in free water of either wells or flowlines may be determined with Hammerschmidt's equation or HYDOFF, as Secs. 2.6 and 2.7 illustrate. Typically the MeOH concentration in the free water of an onshore pipeline is approximately 20 wt%, while offshore MeOH concentrations can exceed 50 wt% if the pressure is high.

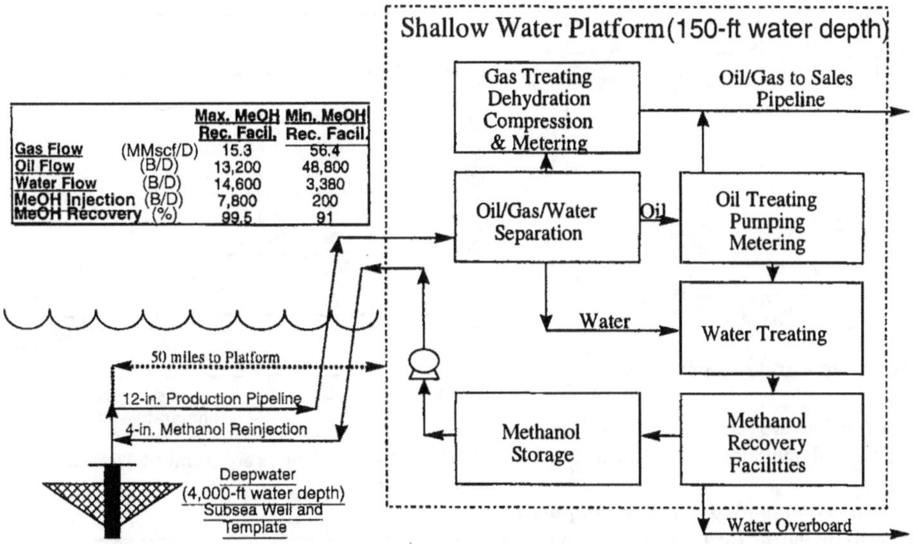

Fig. 2.30—MeOH recovery and injection.[6] (Rec. Facil. = Recovery Facility; Max. = Maximum; Min. = Minimum.) Source: F.S. Manning & R.E. Thompson, Oilfield Processing of Petroleum, Vol.I: Natural Gas Copyright PennWell Books, 1991.

TABLE 2.11—MeOH 1994 COSTS WITH OFFSHORE PLATFORM RECOVERY

Case	High Water (Late Life)		Low Water (Early Life)	
Amount of MeOH in H_2O, wt%	30	20	30	20
Oil production, B/D	13,188	13,191	48,813	48,802
Produced water, B/D	14,593	14,498	3,196	3,188
Gas, MMscf/D	15	15	57	57
Injected MeOH, B/D	7,797	4,537	1,668	960
MeOH loss to gas, lbm/D	6,412	4,430	23,443	15,977
MeOH loss to gas, %	0.29	0.35	4.72	5.53
MeOH loss to oil, lbm/D	2,176	1,456	7,914	5,224
MeOH loss to oil, %	0.10	0.11	1.59	1.81
		Capital and Operating Cost Calculations		
Installed cost on platform, U.S. million $	16.7	13.3	5.03	4.20
Total cost with platform, U.S. million $	20.8	17.4	9.14	8.31
Operating cost, U.S. million $/yr	5.78	4.39	4.25	2.99

A recent finding is that underinhibition with MeOH is worse than no inhibition for two reasons, as measured by Yousif *et al.*[25]: (1) underinhibited systems form hydrates faster than systems without inhibitors and (2) hydrates stick to the pipe walls more aggressively when insufficient MeOH is injected.

While hydrate inhibition occurs in the water phase, significant amounts of MeOH are also dissolved in the vapor and oil/condensate phases. Proportions of MeOH dissolved in the vapor or oil/condensate phases are calculated with the methods in Secs. 2.6 and 2.7 and usually are taken as operating expense losses. MeOH-loss costs can be substantial when the total fraction of either the vapor or the oil/condensate phase is very large relative to the water phase. Case Study 2.2 and Chap. 4 provide sample economics for MeOH.

Makogon[26] noted that, in 1972, the Soviet gas industry used 0.3 kg MeOH/1,000 m^3 gas extracted. Norsk Hydro workers indicated that North Sea MeOH usage may surpass the ratio given by Makogon by an order of magnitude.[27] The use of MeOH in the North Sea has become so expensive that alternatives to MeOH injection are considered.

Case Study 2.2. MeOH Recovery from the Water Phase. Paragon Engineering performed a study for DeepStar on the impact of MeOH recovery on offshore systems.[28] As an evaluation scenario, a conventional, shallow-water platform was designed solely for MeOH recovery in 100 to 150 ft of water, with MeOH return lines 40 to 60 miles to deepwater subsea wells. **Fig. 2.30** shows a block-flow diagram for MeOH recovery and injection.

Costs were determined for MeOH recovery on the platform for eight cases of MeOH in the produced water. **Table 2.11** shows results for four cases: 20 and 30 wt% MeOH in the free water phase for high water production in late field life and for low water production in early field life.

In all cases, MeOH was recovered as the overhead product from a 40-tray distillation column. The bottoms MeOH concentration was less than 1,000 ppm so the water could be dumped overboard.

Paragon Engineering also reported MeOH losses that were greater than anticipated in North Sea recovery systems from three Conoco facilities (4 to 5 gal MeOH lost/MMscf), a Norsk Hydro recovery unit (29% MeOH losses), and an Amoco Netherlands recovery unit (12% MeOH losses).

MeOH recovery is possible from the vapor phase by use of a cryogenic recovery process, but this is seldom done because of expense and safety concerns. MeOH recovery from condensate can be achieved by water wash and subsequent distillation, but this is also seldom done. Environmental and safety (flammability) concerns have a major impact on MeOH recovery.

MEG. Of the glycols, MEG dominates pipeline injection over diethylene glycol (DEG) and TEG because MEG has a lower viscosity and is more effective per pound. MEG also has a higher molecular weight and a lower volatility than MeOH, so MEG may be recovered and recycled more easily on platforms. In addition, MEG losses to the vapor and oil/condensate phases are very small relative to MeOH. Consequently, MEG is most applicable for small water fractions

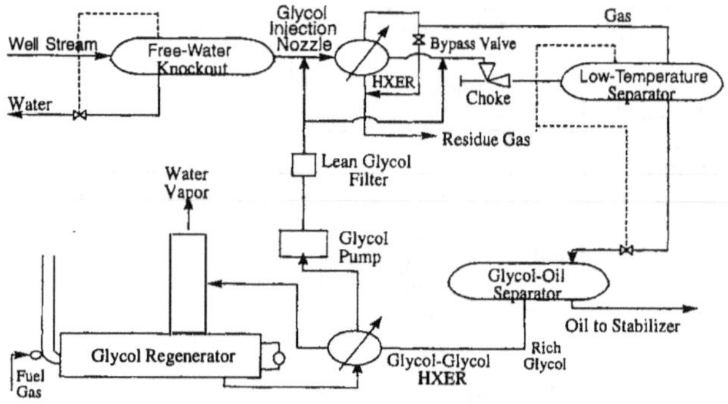

Fig. 2.31—MEG recovery and regeneration.[6]

when gas and oil/condensate fractions are very high. The MEG injection amount may be calculated with methods in Secs. 2.6 and 2.7.

Rule-of-Thumb 10. MEG injection is used when the required MeOH injection rate exceeds 30 gal/hr for onshore gas lines.[6]

Unlike MeOH, MEG's low vapor pressure requires that it be atomized into a pipeline. After injection, MEG is retained with the water phase and provides no hydrate protection above the water level. Because of its high viscosity and density, MEG is seldom used to dissociate a hydrate plug unless the injection point is vertically above a hydrate plug (as in a riser or a well); MeOH is normally used for flowline plugs.

Fig. 2.31 shows a MEG recovery unit, similar to the MeOH recovery block diagram in Fig. 2.30. However, in the MeOH column, the overhead may be almost pure MeOH, while in the glycol regenerator MEG is recovered with water (typically at 60 to 80 wt%) at the bottom. Salt also concentrates in MEG regenerator bottoms (because of low salt vapor pressure) when salt water is produced in the well stream inhibited by MEG. The salt solubility limit in MEG is frequently exceeded, resulting in salt precipitation and fouling of exchangers and other equipment.

Comparison of MeOH and Glycol Injection. In a comprehensive set of experimental studies, Ng *et al.*[29] determined that MeOH inhibited hydrate formation more than an equivalent mass of glycol in the aqueous liquid.

MeOH usage (principally in flowlines and topside on platforms) predominates in the Norwegian sector of the North Sea; MEG is used primarily for hydrates in wells and risers. In contrast, MEG dominates in the U.K. sector of the North Sea. Major problems with use of MEG are high viscosity in long lines and salt precipitation on regeneration. MeOH use is much more prevalent than MEG in the U.S.

Davalath and Barker[3] give a comparison of types of inhibitor injection in a ¼-in. line from surface to subsea tree to prevent hydrate formation in wells. **Fig. 2.32** shows the required quantity of ethylene glycol to be slightly greater than that of MeOH as a function of well temperature. Note the required quantity of TEG is significantly higher than either other inhibitor. **Fig. 2.33** indicates relative pressure drops in the well-inhibitor injection lines, with pump-discharge pressures at 1.8 gal/min for TEG, DEG, MEG, and MeOH of 15,000, 11,500, 7,500, and 4,000 psi, respectively.

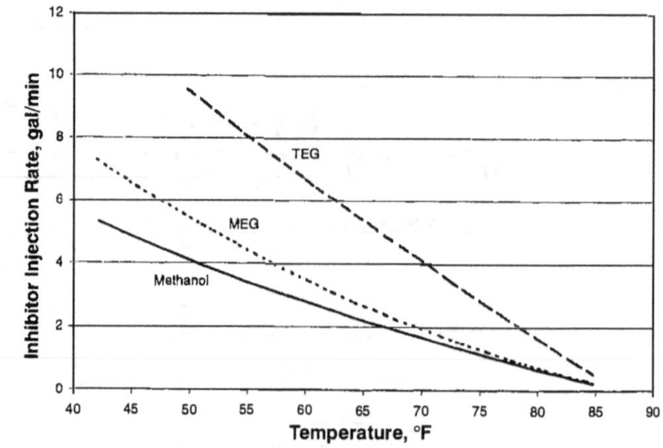

Fig. 2.32—Inhibitor requirement as a function of well temperature.[3]

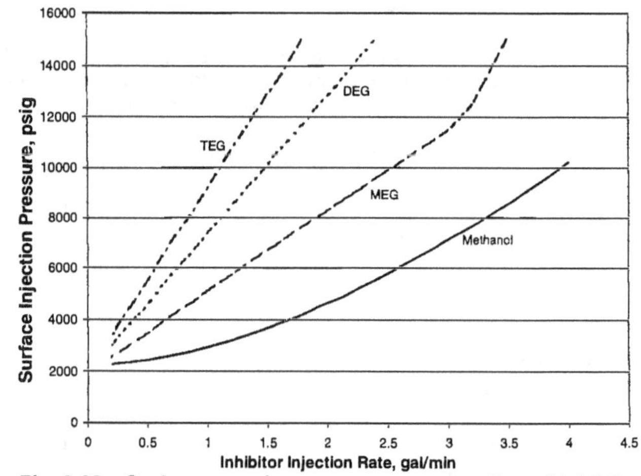

Fig. 2.33—Surface pumping pressure as a function of inhibitor injection rate.[3]

While there is no robust strategy to discriminate between the use of MeOH and MEG to remove a plug, the choice seems to depend on (1) plug location, (2) pressure drop and line size, and (3) properties of the plug in question. **Table 2.12** provides a comparison summary for the general use of the two inhibitors.

Sec. 2.9.3, Guidelines for Use of Kinetic or Thermodynamic Inhibitors, provides a step-wise list of considerations before injection of MeOH and MEG.

2.9.2 Kinetic Control by Antiagglomerants and Kinetic Inhibitors. See Sec. 3.3 of Ref. 8 for the theory of hydrate prevention with the two new techniques of antiagglomeration and kinetic inhibition. At the time of this writing, the kinetic inhibition area is changing rapidly with substantial R&D, and only a few good examples of commercial application exist. With some inhibitors, substantial advantages are claimed for combinations of antiagglomerants and kinetic inhibition.

TABLE 2.12—MeOH AND MEG ATTRIBUTES COMPARISON

	Hydrate Inhibitor	
	MeOH	MEG
Advantages	Easily vaporized into gas	Easy to recover
	For flowline and topside plugs	For plugs in wells and risers
	No salt problems	Low gas and condensate solubility
Disadvantages	Costly to recover	High viscosity inhibits flow
	High gas and condensate losses	Salt precipitation and fouling
	Too little is worse than none	Remains in aqueous phase
	Costly in condensate product (See Table 4.3)	May exsolve unacceptably large amounts of hydrocarbon

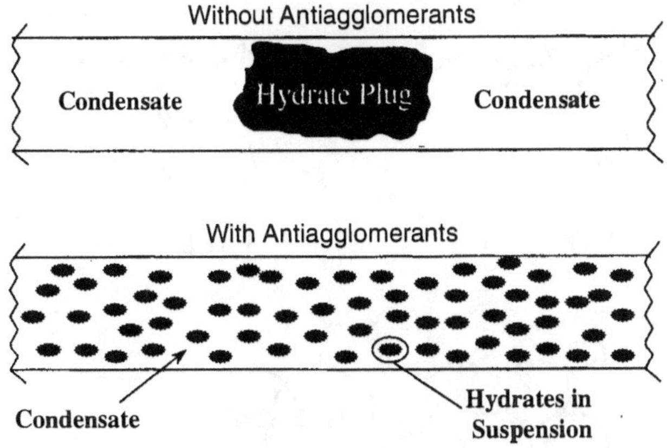

Fig. 2.34—Antiagglomerants in pipeline.[8]

Antiagglomerants. **Fig. 2.34** shows a schematic of the method for antiagglomeration. In the upper diagram, hydrates form large masses and can grow to a size to plug the pipeline. In the lower portion of Fig. 2.34, a dispersant/emulsifier has been added to the gas-condensate system to cause the water to be suspended as small droplets in the condensate.

With this inhibition mechanism, both gas and water are converted to hydrate droplets, but hydrates are prevented from agglomerating to larger hydrate masses capable of plugging pipelines. Even though hydrates are formed, their suspension may provide acceptable flow properties, such as low pressure drops. Antiagglomerant inhibitors are particularly effective in preventing hydrate plugs or flow stoppages, such as shut-ins, with subsequent cooling and restarting.

As dispersant molecules, antiagglomerants have one water-attractive end, while the other end attracts oil, causing a lower surface tension between oil and water. With excess oil, a surfactant causes the water phase to be suspended as emulsified droplets. However, with excess water, the emulsion may be reversed and water will be the external phase. Surfactant chemistry is complex, and a different surfactant may be required to emulsify water with each oil (or condensate).

Lingelem *et al.*[30] presented the Norsk Hydro data in **Fig. 2.35** as an example of antiagglomerant behavior in a multiphase pipeline that did not exhibit plugging when the initial water/oil ratio (WOR) was less than 60 vol%, even with hydrate formation. In contrast, plugging was observed above 60 vol% WOR with less than 10 wt% MeOH in the free water. Other multiphase oil pipelines (with fluids not shown in Fig. 2.35) commonly plug with minimal WOR.

The Norsk Hydro authors suggest that behavior like that in Fig. 2.35 illustrates a "natural" antiagglomerant mechanism because "the difference in plugging behavior is attributed to the type and amount of natural surfactants present in the oil or condensate. In general, oils with little tendency to form stable emulsions have been observed to form hydrate plugs more easily than oils more prone to form stable emulsions."

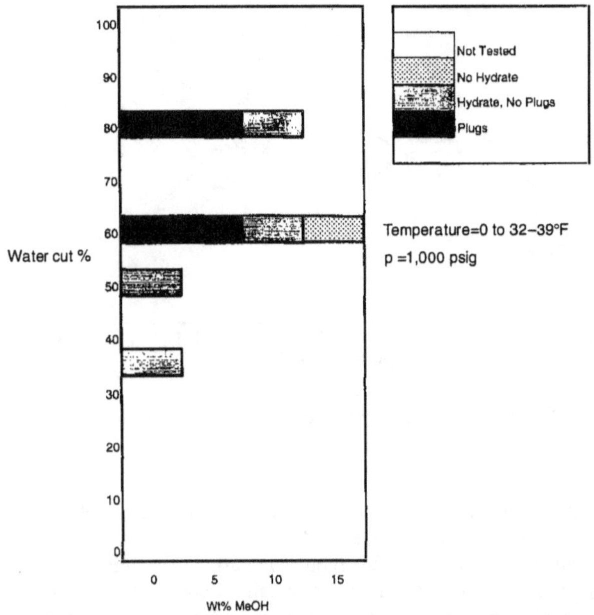

Fig. 2.35—Antiagglomerants effectiveness in various amounts of water.[30]

Rule-of-Thumb 11. Use of antiagglomerants requires a substantial oil/condensate phase. The maximum WOR (volume basis) for the use of an antiagglomerant is 40 : 60.

This rule of thumb is based on two observations.

1. At WOR higher than 40 : 60, the water-in-oil emulsion may invert to an oil-in-water emulsion. If the water phase is external, hydrates will grow beyond small droplets.
2. Coal-slurry-transport technology provides a maximum ratio of coal : liquid vehicle of 40 : 60. Higher ratios increase the risk of having a nontransportable hydrate phase, similar to a nontransportable coal slurry.

At the French Petroleum Inst. (IFP), Behar *et al.*[31] provided three performance examples of an antiagglomerant inhibitor in a 2-in. pilot loop for a recombined crude, a gas-saturated oil, and a gas-saturated condensate with WOR's of 0.3, 0.3, and 0.1 ft^3/ft^3, respectively. The condensate example is shown without and with anti-agglomerant inhibition in **Figs. 2.36 and 2.37,** respectively.

Gas consumption increases in both cases, indicating hydrate formation even with antiagglomerant inhibition. However, the loop pressure drop (an indicator of a system without plugging) remains at a low value with inhibition (Fig. 2.37), while it increases rapidly with

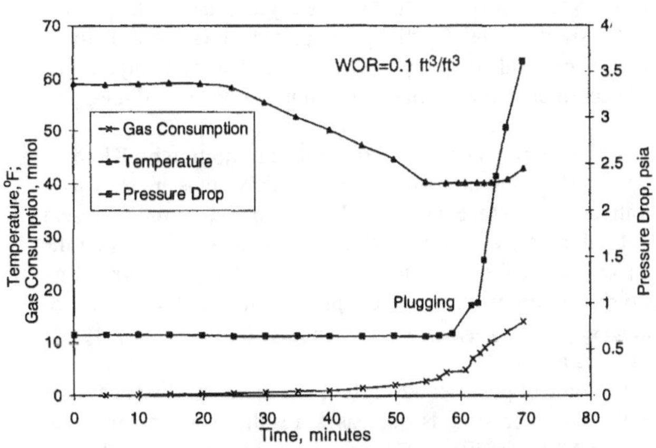

Fig. 2.36—Hydrate formation with plugging.[31]

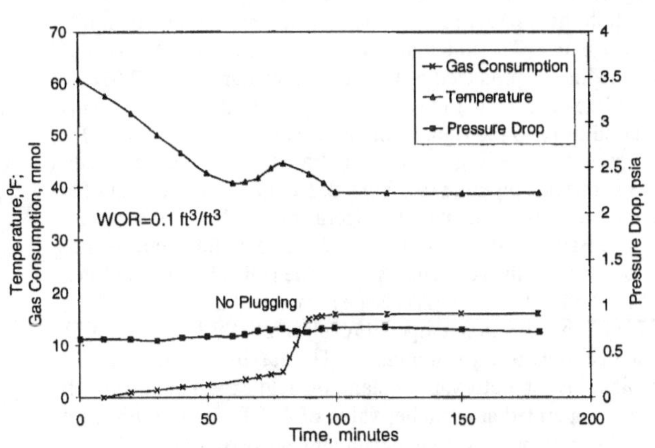

Fig. 2.37—Antiagglomerants preventing plugging.[31]

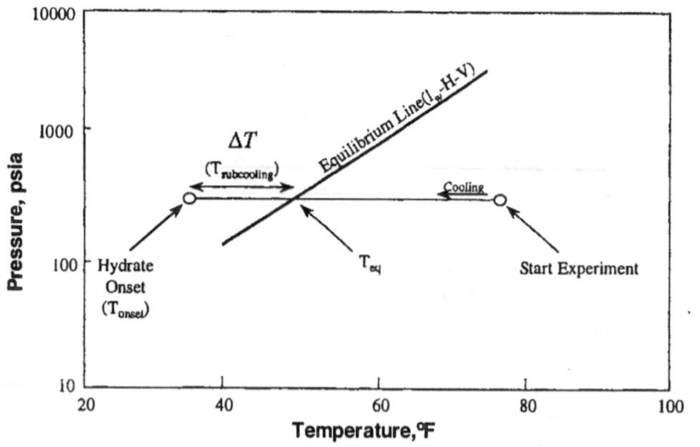

Fig. 2.38—Subcooling as a measure of kinetic inhibitors.

out inhibitor in Fig. 2.36. A low pressure drop indicates that the effective viscosity is small and that the fluid components flow readily.

Small amounts of surfactant are required relative to traditional inhibitors like MeOH. Behar *et al.*[31] indicate that 1 wt% of an emulsifier is equivalent to 25 wt% MeOH. Economics should include such factors as surfactant cost, emulsion breaking, and recovery and environmental considerations.

Specific surfactants or dispersants must be formulated and tested as emulsifying agents for each composition of condensate. Surfactant choice is not straightforward; many surfactants have been shown to promote hydrate formation. Significant technology was transferred from earlier studies of enhanced oil recovery.

Undocumented reports from Shell report an inhibition chemical that provides inhibition at an order of magnitude lower concentration than the IFP chemical without being condensate specific. Reportedly, this additive allows the hydrates to form before taking them into the condensate phase; however, some environmental concerns persist.

There is no published commercial example of the use of an antiagglomerant in an offshore hydrates application. Yet the method holds great promise, especially for deep, highly subcooled systems and shutdown with cold-restart situations. Weaknesses of the method include toxicity concerns, the need to break emulsions, and the need to recover the expensive dispersant additive. Antiagglomerant chemicals are proprietary and chemical structures, properties, and performance are not in the open literature. The next decade will undoubtedly see major advances in these chemicals.

Kinetic Inhibition. Kinetic inhibition of hydrate growth has a different mechanism than that of antiagglomerants. While there is evidence that the presence of a liquid hydrocarbon phase aids inhibition, kinetic inhibitors prevent hydrate-crystal nucleation and growth without the requirement of a liquid hydrocarbon phase. Nucleation inhibition prevents hydrate crystals from growing to a critical radius. Growth inhibition maintains small hydrate crystals, inhibiting growth of larger crystal masses.

Fig. 2.38 shows ΔT, the most common measure of kinetic inhibitor performance. The line marked L_w-H-V represents the hydrate-formation line, as predicted by the gas-gravity curve (Sec. 2.6) or by HYDOFF (Sec. 2.7), with hydrate formation to the left and nonhydrate conditions to the right of the line. The horizontal line in Fig. 2.38 represents a cooling curve for hydrate-forming mixtures, such as may occur in a pipeline (see Figs. 2.2 and 2.3). The object of kinetic inhibition is to maintain the operating condition of a pipeline as far as possible to the left of the L_w-H-V line without hydrate plug formation during the residence time of the fluids in the flowline.

In Fig. 2.38, subcooling (ΔT) is the measure of the lowest, nonhydrating system operation temperature relative to the hydrate equilibrium temperature at a given pressure. The maximum value of ΔT is determined by laboratory, pilot plant, or field experiments, and the pipeline is operated at a smaller value of ΔT. The maximum value of ΔT appears to be pressure independent; however, ΔT_{max} does depend on the polymer; molecular weight; and the amount of salt, glycol, and alcohol present.

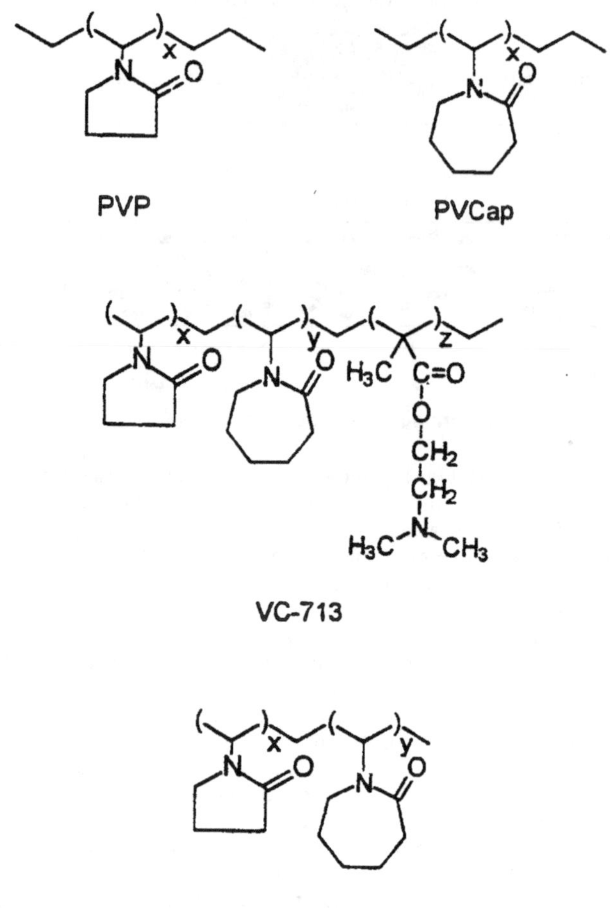

Fig. 2.39—Formulas of some kinetic inhibitors.[8]

Because the subcooling temperature is the difference between the equilibrium temperature, T_{eq}, and the operating temperature, T_{op}, a decrease in subcooling can be obtained by either increasing the lowest T_{op} or by decreasing T_{eq} through the use of MeOH or glycols. Thus, thermodynamic inhibitors can be combined with kinetic inhibitors to achieve the desired subcooling. Recent measurements indicate water residence time can be longer than 30 days without hydrate formation, when the lowest pipeline ΔT is at a temperature at least 3°F less than the maximum subcooling, ΔT_{max}, with a good kinetic inhibitor.[8]

Kinetic inhibitors are commonly polymers with several chemical formulas shown in **Fig. 2.39**. Each chemical shown has a polyethylene backbone connected to pendant groups containing an amide (-N-C=O) linkage, frequently within a five- or seven-member ring. As the inhibitor adsorbs on the hydrate crystal, the pendant group penetrates specific sites (partially completed cages) on a hydrate-crystal surface, while the polymer backbone extends along the surface. To continue growing, the crystal must grow around the polymer; otherwise, crystal growth is blocked.

Fig. 2.40 shows a mechanism for kinetic inhibition by Klomp.* Adsorption of three kinetic-inhibitor polymer strands are shown on a hydrate-crystal surface. The "filled stars" on each strand represent pendant polymer groups that dock at the "empty-star" sites on the hydrate crystal surface. As indicated in the figure, the subcooling ΔT is directly proportional to the liquid-crystal surface tension, σ, but inversely proportional to the length, L, between polymer strands; k_l is a constant.

If the amount of polymer adsorption increases, the distance L between the polymer strands decreases, resulting in an increased subcooling ΔT performance. Conversely, if the amount of inhibitor adsorption decreases (caused by depletion by multiple small hy-

*Personal communication with U. Klomp, Royal Dutch Shell, Amsterdam (5 June 1996).

$$\Delta T \leq \frac{4\sigma}{k_I \cdot L}$$

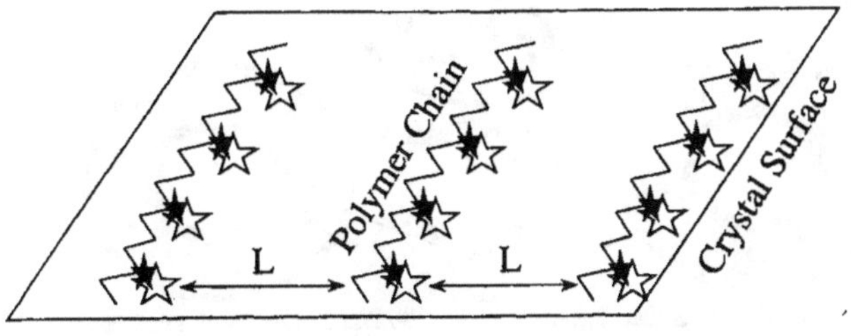

Fig. 2.40—Polymer adsorption in hydrate crystal.[8]

drate crystals) the distance L between polymer strands increases, resulting in a smaller subcooling ΔT.

One of the first kinetic inhibitors developed was polyvinylpyrrolidone (PVP), a polymer with the structure shown in Fig. 2.39. Several companies have adopted the use of PVP in onshore fields with a small subcooling (ΔT) and short residence time.

Arco[32] and Texaco[33] reported initial field tests of kinetic inhibitors. Bloys[32] reported the effectiveness of 0.3 to 0.4 wt% VC-713 in a 17 day test in a North Sea pipeline. The pipeline (8-in. diameter, 5.8 miles long) had a flow rate of 20 MMscf/D, 0.5 B/MMscf condensate, and 0.2 B/MMscf condensed water at a subcooling between 2°F and 16°F. Bloys suggested that economics were very favorable for new developments but marginal for retrofits of systems with traditional inhibitors such as MEG.

Notz *et al.*[33] indicated the successful, routine use of PVP in several wells and flowlines in Texas and in Wyoming. Notz further reported that PVP (at less than 1 wt%) was effective in replacing MeOH (at concentrations from 10 to 60 wt% in free water), resulting in savings of as much as 50%. See Texaco Case Studies C.13 and C.14 in Appendix C.

Rule-of-Thumb 12. PVP may be used to inhibit pipelines with subcooling less than 10°F for flowlines with short gas residence times (less than 20 minutes).

Rule-of-Thumb 12 comes from Ref. 7, which reflects Texaco operating kinetic-inhibitor practice with approximately 30 flowlines from their Brookeland field.

Other kinetic inhibitors are more effective (but more expensive per pound) than PVP (illustrated by the other chemical formulas in Fig. 2.39) and have a seven-member ring pendant group in place of the five-member PVP pendant ring. The better kinetic inhibitors provide additional subcooling with long water residence times.

Rule-of-Thumb 13. VC-713, PVCap, and copolymers of PVCap can be used to inhibit flowlines at subcooling less than 18°F, with water-phase residence times up to 30 days.

Rule-of-Thumb 13 comes from commercial use of kinetic inhibitors, as indicated in the next case study.

Case Study 2.3: North Sea Use of New Inhibitors. On 22 July 1996, British Petroleum (BP) initiated continuous commercial use of kinetic inhibitors [called threshold hydrate inhibitors, (THI)] in flowlines in the West Sole gas-export lines.[34] This followed from an extensive set of field trials carried out in the Ravenspurn to Cleeton wet-gas line.[35]

BP began THI use in a 16-in.-inner-diameter (ID), 13-mile long pipeline from three Ravenspurn platforms to Cleeton. At the time of the trial the maximum flow rate was 195 MMscf/D. For the purpose of the trial the flow rate was cut back to 90 MMscf/D to put the line as far into the hydrate region as possible (16°F of subcooling). Three trials were carried out, all of which were successful. The trials included extensive periods of shut-in, up to 7 days with successful restart. A typical water-production rate was 1.6 B/MMscf with a line pressure of 1,088 psia and a low temperature of 48°F. The THI dose rate was 3,000 to 5,000 ppm on the basis of the free-water phase. See Ref. 35 for further details.

Two lines [24- and 16-in. ID, 35 miles long] from West Sole A, B, and C and from Hyde have been inhibited with THI for more than 2 years. Water residence times can be as long as 2 to 3 weeks, and the lines are 11°F inside the hydrate region at operational conditions. The gas is very lean, producing very little condensate. Water content is low, and free water comes mainly from condensation. Because of the low amounts of water and condensate, this is an atypical case, but nonetheless represents a commercial test of kinetic inhibitors.

Water production from all four West Sole platforms is 150 to 200 B/D, or approximately 0.3 bbl/MMscf with 250-MMscf/D total produced from the three West Sole platforms and the remainder from the Hyde platform. The condensate rate is also 0.3 bbl/MMscf. The THI pumping rate per platform is 2 to 3 L/hr of solution that contains approximately 15 wt% active ingredient. The target injection rate is 3,000 ppm based on the free-water phase.

2.9.3. Guidelines for Use of Chemical Inhibitors. The next section provides a stepwise protocol to determine whether use of inhibitors might be suitable, modified from the original suggestions by Edwards* and T.R. Oil Services.** It should be emphasized that, before field application, experimental data should be obtained, particularly if kinetic inhibition is being considered. The protocol provides preliminary steps for such experiments.

Guidelines for Use of Kinetic or Thermodynamic Inhibitors.

1. If the field is mature, record the current hydrate-prevention strategy. Record the existing and planned procedures for dealing with an unplanned shutdown. Provide a generic description of the chemistry of the scale and corrosion inhibitors used.
2. Obtain accurate gas, condensate, and water analyses during a field drill test. Estimate how these compositions will change over the life of the field. Estimate the production rates of gas, oil, and water phases over the life of the field.
3. Generate the hydrate pressure/temperature equilibrium line with several prediction methods. If the operating conditions are close to the hydrate line, confirm the prediction with experiment(s).
4. Determine the water-production profile over field life (see the next subsection, One Scenario for Pipeline Water Over Field Life, for an example).
5. Consider the pipeline topography along the ocean floor to determine where water accumulations will occur at dips, resulting in points of possible hydrate formation.
6. Simulate the pipeline pressure/temperature profile with a simulator, such as OLGA or Pipesim®, to perform hydraulic and heat-transfer calculations in the well, flowlines, and separator over the life of the field.
7. Determine the water-residence times in all parts of the system, especially in low points of the pipeline.
8. Estimate the subcooling ΔT (at the lowest temperature and highest pressure) relative to the equilibrium line over all parts of the system, including fluid separators and water-handling facilities. List the parts of the system that require protection.
9. If ΔT<14°F, consider the use of kinetic inhibitors. If ΔT>14°F, consider the use of standard thermodynamic inhibitors possibly combined with kinetic inhibitors or antiagglomerants (no one has used antiagglomerants commercially as of 1 November 1999).
10. Perform economic calculations (capital and operating expenses) for four options (1) drying, (2) MeOH, (3) MEG, and (4) kinetic inhibitors.
11. Determine whether inhibitor recovery is economical.
12. Design the system hardware to measure (1) temperature and pressure at pipe inlet and outlet, (2) water monitor for rates at receiving facility, and (3) the following chemical check list.
   - Has the inhibitor been tested with systems at the pipeline temperature and pressure?

*Personal communication with A.R. Edwards, BP Dimlington Terminal, U.K. (17 July 1997).

**Personal communication with M. Grainger, "The Oil Services Perspectives on the Use of Alternate Hydrate Inhibitors,"TR Oil Services Inc., Aberdeen, U.K. (3 September 1997).

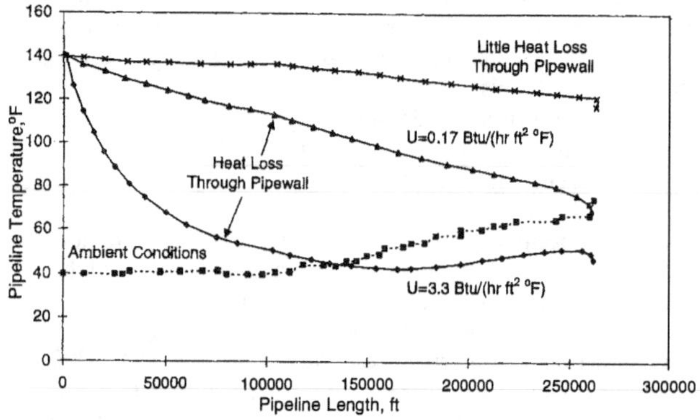

Fig. 2.41—Pipeline temperature vs. change in heat conductivity of pipeline.[40]

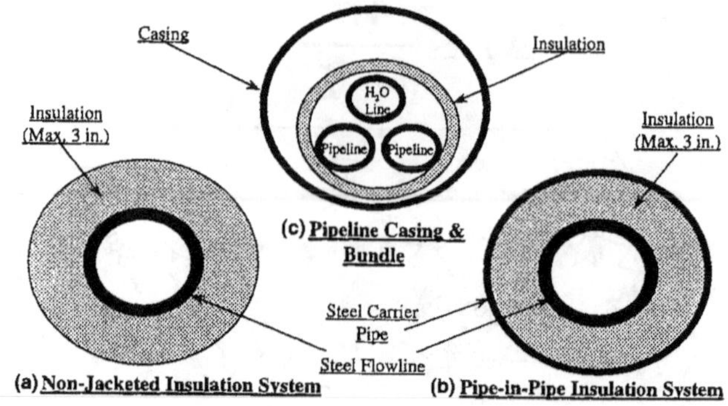

Fig. 2.42—Pipeline insulation methods.[36]

- Consider the environmental, safety, and health impacts of the chemical.
- Determine physical properties, such as flash point (should be $>135°F$), viscosity (should be <200 cp at lowest T), density, and pour point (should be $>15°F$).
- Determine the minimum, maximum, and average dosage of inhibitor.
- Determine the storage and injection deployment methods.
- Determine the material compatibility with gaskets, seals, and other such equipment.
- Determine compatibility with other production chemicals.
- Determine the compatibility with the process downstream including cloud point, foaming, and emulsification tendencies.

At an early stage in the inhibitor design process, it will be worthwhile to consider obtaining laboratory data and involving a service company to provide field support of process hydrate inhibition.

Hydrate inhibition occurs in the water phase and is dependent on the amount of water production and the salt concentration. Because the amount and concentration of pipeline water depends heavily on produced water, reservoir engineers should provide the best estimate of produced water and salt concentrations over the life of the field. The second source of water, condensed water, can be estimated from gas water content as illustrated in the Amount of Water Phase section in Sec. 2.6.3. and in Fig. 2.9.

Edwards* provided one possible scenario for water production over a field life, given in the next section. The next scenario for a North Sea pipeline indicates a nonintuitive need for inhibition. The pipeline initially operates with a low inhibition need; in midlife the pipeline requires inhibition; and in the final stages of life, the pipeline does not require hydrate inhibition.

One Scenario for Pipeline Water Over Field Life.

1. At the beginning of field life, water production may be low so that only a small amount of condensed water can be responsible for hydrate problems. The field may be operating with a large subcooling ΔT, but low dosages of chemicals are required because of low amount of water production. However, there are counter-examples of fields that begin producing water early in their life. Only fluid measurements can assess this difference.

2. At field midlife, water produced down the line (if there is no upstream separation facilities) increases. Both produced water and condensed water may be substantial. Total water may double or triple, but the condensed-water amount may be sufficient to dilute the solution to low salt concentrations, so that maximum inhibitor injection rates may be required. Over a field lifetime, typical salt concentration from produced water may vary from 0% to nearly the reservoir concentration.

3. At the end of field life, there may be as much as 10 times as much water, but it is mostly saline production. Both the increase in water salinity and the pressure decline of the field may take the production fluids outside the hydrate pressure/temperature region.

4. As an example, one BP field is forecast to dip into and out of the hydrate-formation region over its life.

*Personal communication with A.R. Edwards, BP Dimlington Terminal, U.K. (17 July 1997).

2.9.4 Heat Management. The discussion in this section has been excerpted from DeepStar reports,[36-40] from Aarseth,[41] and from discussions with Statoil researchers.

Retention of reservoir heat is one of the most efficient means of hydrate prevention. Because all reservoirs contain water and because water acts as a heat sink owing to a high heat of vaporization, fluids at the wellhead are typically at temperatures from 140 to 212°F. When the reservoir fluid flows through cool ambient surroundings, the temperature can cool quickly into the hydrate region as determined by the heat-transfer coefficient, U, between pipe and surroundings.

Case Study 2.4. Pipeline Temperature With Heat Loss. Fig. 2.41 shows an offshore pipeline temperature as a function of length at various values of U between the pipeline inlet temperature (140°F) and a separator 50 miles away.[40] The pipeline in the figure was assumed to be 50% buried and had a water flow of 1,527 B/D, a gas flow of 30.76 MMscf/D, and an oil flow of 22,723 B/D.

In Fig. 2.41, the lowest dashed line shows the temperature of the ocean with pipeline length. The upper lines represent the pipeline temperature with no heat loss through the pipe wall, with the temperature drop resulting from expansion. However, with heat loss through the pipe wall, the temperature drops dramatically. With overall heat-transfer coefficients of $U=0.17$ and $U=3.3$ Btu/hr-ft^2-°F, the separator temperature is 70 or 50°F respectively, where hydrates can easily form. It is not clear why the lower line decreases below the ambient condition at the pipeline end (perhaps owing to expansion). However, Fig. 2.41 does indicate that a lower heat-transfer coefficient provides higher temperatures needed to prevent hydrate formation.

The temperature profiles in this case study are for a flowing pipeline. If the pipeline were shut-in, the system would cool rapidly to the ambient conditions represented by the dashed line at the bottom of Fig. 2.41. At the low ambient temperatures, hydrate problems are particularly severe and blockages may occur, particularly when the system is restarted.

If hydrates form in an insulated pipeline, the pipeline can be depressurized to achieve a low hydrate equilibrium temperature, so that heat can flow into the hydrates from the surroundings, which are at higher temperatures. In such cases, insulation decreases heat flow from the surroundings, making hydrate dissociation much more difficult. As a consequence, it is good operating practice to "pig" the line and inject large quantities of MeOH or MEG into the pipeline before a planned shut-in.

Hydrates can be prevented in pipelines by three types of heat control:

1. Burying the pipeline to provide heating and insulation by the ocean floor.
2. Insulating the pipeline with nonjacketed insulation, pipe-in-pipe (PIP) systems, and bundling systems.
3. Heating the pipeline.

Pipeline burial is a good means of providing pipeline insulation and protection. The degree of insulation depends on the thermal gradient in the earth along the pipeline route, the pipeline depth, and the ambient temperature. Expenses for providing a trench and burial

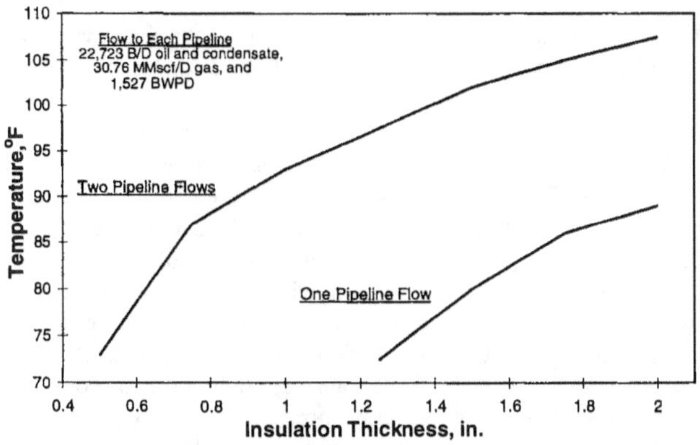

Fig. 2.43—Riser temperature vs. thickness.[40]

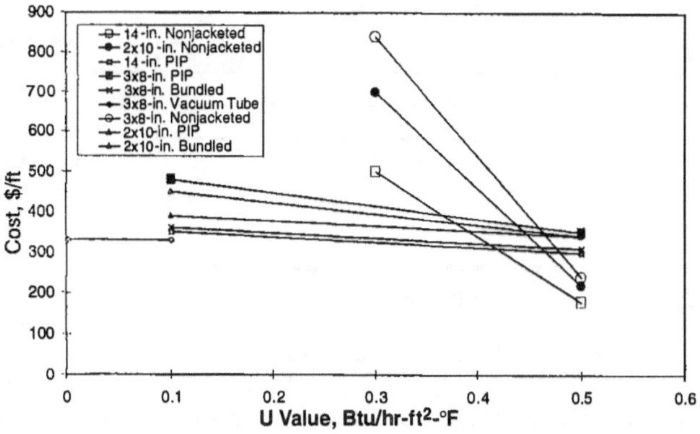

Fig. 2.44—Cost vs. overall heat-transfer coefficient (depth, 6,000 ft, production rate, 25,000 BOPD).[36]

system for pipelines may be very high, particularly at great ocean depths. On the other hand, heat-control systems through pipeline insulation or heating may be installed with the pipeline. Pipeline insulation and heating methods are given consideration in design, but insulation alone offers no protection for long-term shut-ins.

Insulation Methods. **Fig. 2.42** shows the three categories of insulated pipelines: nonjacketed, PIP, and bundled flowlines. The nonjacketed system (Fig. 2.42a) consists of an insulated pipe with a coating. The minimum overall coefficient achievable with a nonjacketed system is 0.3 Btu/hr-ft^2-°F and costs are typically U.S. \$50 to \$300/ft (1995 dollars) for pipes with diameters between 8 and 12 in.[36]

The PIP system (Fig. 2.42b) is the most thoroughly tested of the three types. In this system, the flow pipe is within an outer pipe, with either insulation or vacuum between the two pipes, sometimes aided by a reflecting screen in the annulus. With a 3- to 4-in. insulation layer, the PIP system can provide an overall heat-transfer coefficient, U, of 0.14 to 0.6 Btu/hr-ft^2-°F.

Fig. 2.42c shows a bundled line with two or more flowlines and a startup water line with an insulator, all in an outer casing. For offshore applications, bundles are fabricated onshore in lengths up to 10 miles and towed to their offshore position, currently at water depths of up to 5,000 ft. Overall heat-transfer coefficients as low as 0.1 Btu/hr-ft^2-°F can be achieved.

Fig. 2.43 shows the increase in temperature at the platform riser as a function of insulation thickness, with two pipelines flowing together compared with an individual flow, when each line has a water flow of 1,527 B/D, a gas flow of 30.76 MM scf/D, and an oil flow of 22,723 B/D.[10] The addition of a second flowline can reduce the insulation thickness required to obtain a given riser temperature, or the second flowline can provide a higher riser temperature for a given insulation.

Figs. 2.44 and 2.45 compare the cost (in 1995 dollars) of the three types of insulation for water depths of 6,000 ft over 60 miles at oil-production rates of 25,000 and 50,000 B/D, respectively.[36] If an average $U=0.3$ Btu/hr-ft^2-°F is required with a flowline pressure of 4,000 psia, bundled flowlines are more cost-effective.

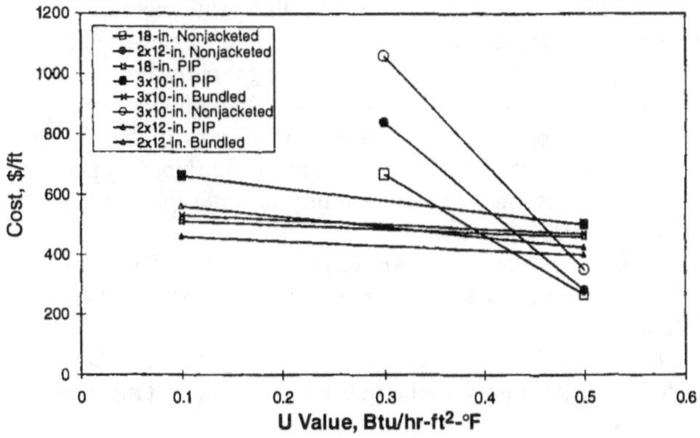

Fig. 2.45—Cost vs. overall heat-transfer coefficient (depth, 6,000 ft, production rate, 50,000 BOPD).[36]

Pipeline Heating Methods. DeepStar Report CTR A601-b concludes that "where pipeline depth precludes depressurization below the hydrate-formation pressure, heating may be the only option to clear a hydrate plug."[37] Yet, as noted in Chap. 1, heating a hydrate plug can be very dangerous. DeepStar Report CTR A601-c concludes that pipeline heating will be very expensive, "at least 1MW of power is required for a 36°F increase of a 10-in. pipeline 15 miles long."[38]

In the future, Statoil will use heating more extensively to reduce use of MeOH or other chemicals to prevent hydrates. The Chevron/Conoco Britannia project started in the North Sea in 1998 with a bundled line to heat fluids. The three common means of heat management are (1) bundling hot water lines (for 6.2 miles and return) in production lines to prevent hydrate formation, (2) Combipipe (**Fig. 2.46**) for induction heating, with current flowing through three cables outside of pipe but within insulation, with corrosion protection, and (3) direct electrical heating for 31 to 37 mile long lines (**Fig. 2.47**) in which the pipeline is the primary conductor with a current return line at 3.3 ft in parallel to the pipeline. It should be noted however, that such heating tools are in the planning stage and commercial use has yet to be documented.

2.10 Hydrate Prevention During Drilling and Testing

In most onshore drilling and testing, the reservoir temperature is sufficient to prevent hydrate formation except downstream of large pressure drops, such as the choke. In Arctic and offshore environments, however, low temperatures often cause hydrate formation in both drilling and testing in addition to the hazard of drilling through hydrated reservoirs. The purpose of this section is to enable the engineer to design a drilling fluid to prevent hydrate formation (Sec. 2.10.1), test a well with both gas expansion and cooling without forming hydrates (Sec. 2.10.2), and safely drill through hydrates (Sec. 2.10.3).

2.10.1 Hydrate Prevention in Drilling Fluids. As indicated in Sec. 2.4, the principal hydrate concern in deepwater drilling fluids is formation of plugs in the choke and kill lines. Barker and Gomez[5] first reported hydrate formation resulting from gas kicks during

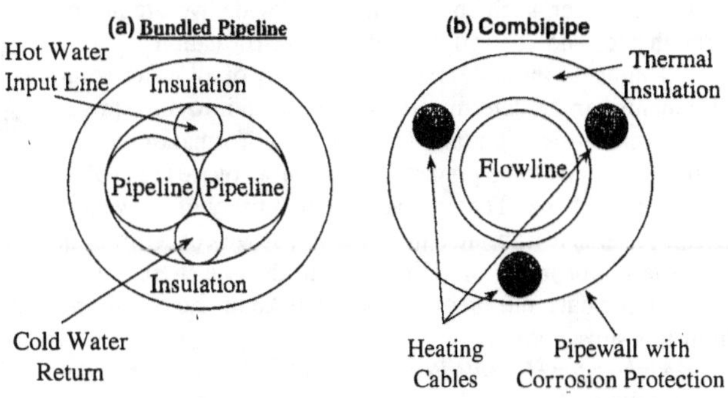

Fig. 2.46—Heating through (a) bundling and (b) combipipe.

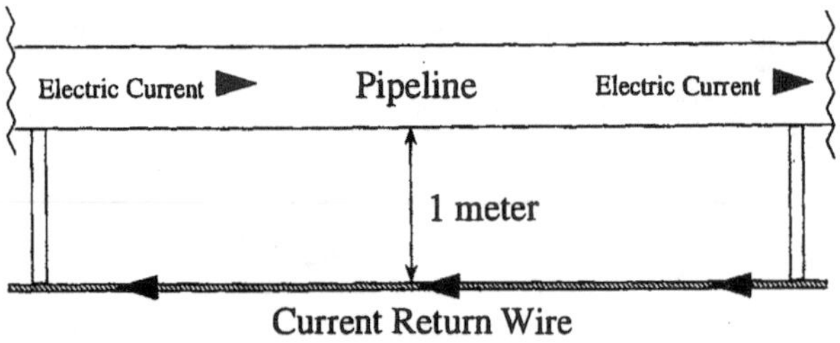

Fig. 2.47—Direct electric heating.

deepwater drilling with two case studies listed in Appendix C—Case Study C21 off California, in which 70 days' drill time was lost, and Case Study C22 in the Gulf of Mexico, in which 50 days' drilling time was lost. In their classic paper, Barker and Gomez summarize four specific adverse affects of hydrates in drilling:

1. Choke and kill line plugging, preventing their use in well circulation.
2. Hydrate-plug formation at or below the BOP, preventing well-pressure monitoring below the BOP.
3. Hydrate-plug formation around the drillstring in the riser, BOP's, or casing, which prevents drillstring movement.
4. Plug formation in the ram cavity of a closed BOP, which prevents the BOP from opening fully.

Synthetic Oil-Based Drilling Fluids. Synthetic oil-based drilling fluids have very low toxicity and good bioremediation qualities—two properties that allow disposal of cuttings offshore. Because of their inherent drilling advantages and hydrate suppression, synthetic oil-based fluids predominate in deepwater drilling. Grigg and Lynes[42] first showed that oil-based drilling fluids contain sufficient water to form hydrates. However, hydrate occurrences are unlikely in oil-based fluids, particularly if the internal brine phase has high salt content (e.g., 25 to 30% $CaCl_2$).

Water-Based Drilling Fluids. It is becoming more difficult to use oil-based drilling fluids offshore with increasingly stringent environmental regulations, thus acceptable water-based fluids are needed. Hale[43] indicated Shell's experience with the use of water-based fluids in deepwater drilling. By 1988, Shell had drilled 16 wells in the Gulf of Mexico at water depths between 2,000 and 7,500 ft by use of 20 wt% sodium chloride (NaCl) and partially hydrolyzed polyacrylamide (PHPA) fluids. Shell experienced an average of more than one gas kick per well, indicating possible hydrate formation.

An industrial consortium funded drilling-fluid experiments at the Colorado School of Mines,[44-46] resulting in a rule of thumb to predict the formation of hydrates in water-based drilling fluids.

Rule-of-Thumb 14. The water and the water-soluble drilling-fluid components determine the conditions of hydrate formation. Concentration of salts, alcohols, glycols, and glycerols in the water determine the hydrate-formation temperature and pressure.

The nonwater-soluble drilling-fluid components (e.g., clays, barite, PHPA) do not affect the pressures and temperatures at which hydrates form. However, such components do provide hydrate nucleation/growth sites and can influence the rate of formation.

Rule-of-Thumb 14 indicates the water activity of all water-soluble components can be determined and translated into the equivalent salt concentration. This allows the hydrate-formation conditions of drilling fluids to be predicted with the software HYDOFF, provided with this book. That is, once water activity of the water-soluble components is translated into equivalent salt concentration, the normal inhibitor mode of HYDOFF can then be used. Determination of input data and prediction steps make up the remaining portion of this subsection.

Input Variables for Hydrate Prediction in Drilling Fluids. If development wells are being drilled in an existing field, the following information may be available. If an exploration well is being drilled, educated guesses are needed for the input variables. In such cases, repetitive calculations should be made to determine the sensitivity of hydrate-formation conditions (and resulting mud compositions) to a range of the next variables.

1. The normal temperature and pressure profiles are needed in the well, both at flowing and at static conditions. Commercial programs, such as WELLTEMP or WEST, may be used for this purpose. The worst case to be considered is usually a long-term shut-in with fluid cooling to seabed temperature at the wellhead.
2. An estimate of gas composition is needed. Consider gas composition of offset wells; if no data are available, assume a conservative high propane content, such as 90% methane, 7% ethane, 3% propane. Hydrate formation is strongly influenced by propane concentration. Consider performing a concentration sensitivity study.
3. Abnormal operation pressures should be considered, such as during a gas kick. Quantifying abnormality is difficult and requires bounding the conditions. In the worst case, it may be assumed that the gas kick displaces all the mud column and that the well experiences full reservoir pressure to surface. A typical kick pressure is 1,000 psia.* The position of greatest jeopardy is where the temperature is lowest, close to the seafloor. Skalle and Podio[47] totaled the Gulf of Mexico Outer Continental Shelf blowouts from 1960 to 1996 in terms of well depths: < 5,000 ft (64 or 57%), 5 to 10,000 ft (18 or 16%), 10 to 15,000 ft (22 or 19%), and > 15,000 ft (9 or 8%). Note that there is a bias in these numbers because most wells were drilled at the lower-depth region.
4. The salt content of produced water should be known because salt provides an additional amount of inhibition. For a conservative answer, freshwater should be assumed.

Commonly Available Water-Based Drilling Fluids. Hale and Dewan[48] showed the effects of salts and glycerol on hydrate formation in deepwater applications. Lai and Dzialowski[49] performed experiments to show that hydrates can form in many drilling fluids at pressures less than 2,000 psig. On a weight basis, NaCl is the most effective hydrate inhibitor, followed in decreasing order by potassium chloride (KCl), calcium chloride ($CaCl_2$), sodium bromide (NaBr), sodium formate (NaCOOH), and calcium nitrate [$Ca(NO_3)_2$].

1. In a recent survey of the drilling literature, Ebeltoft and coworkers[50] reported that the most common deepwater drilling fluids are 20 to 23 wt% NaCl/PHPA systems that have been used successfully at depths up to 7,500 ft. A typical drilling fluid would be composed as follows: bentonite [3 to 5 ppb(lbm/bbl)], Xanthan gum, (0.5 to 1 ppb), PAC LV (2.0 ppb), NaCl (20% or 82 ppb), PHPA (0.5 to 1.0 ppb), soda ash (as needed), barite (as needed), and water (as needed).
2. High-KCl (< 24 wt%) drilling fluids are the second-most-frequently used. While KCl gives very strong shale stability, it is a much weaker hydrate inhibitor than NaCl.
3. There is experience with 18 $CaCl_2$ drilling fluids in the field at concentrations of < 32 wt%. These drilling fluids provide fairly strong shale inhibition and are capable of excellent hydrate inhibition but are difficult to formulate as good drilling fluid because of the interaction of calcium ions (Ca^{++}) with clay and water-soluble polymers. Improved $CaCl_2$ systems are evolving rapidly. It may be difficult for high-$CaCl_2$ drilling fluids to pass Gulf of Mexico bioassay tests; therefore, blends of $CaCl_2$ may be made with high levels of $NaCl_2$ to balance hydrate inhibition with acceptable bioassay performance. $CaCl_2$ has much higher solubility in water than $NaCl_2$, and gives up to 80°F suppression.

Potential Drilling Fluids That Are Rapidly Evolving. Two additional types of drilling fluids are imminent, and field tests should be available shortly.

4. Blends of high NaCl plus low-molecular-weight glycols have been tested in the laboratory. Field trials were imminent at the time of this writing.
5. Formate (potassium formate—K COOH) drilling fluids are being tested that are soluble up to 75 wt% in water. NaCOOH has a

*Personal communication with J.B. Bloys, Arco E&P Technology, Plano, Texas (11 July 1998).

lower solubility, but has not been tested specifically for deepwater applications to date. These are very expensive but can provide substantial hydrate inhibition.

Example 2.14: Hydrate-Formation Conditions With a Drilling Mud. We wish to predict hydrate-forming conditions for a mud to be used in a well to be drilled in 5,000 ft of water where the mudline temperature is 39.8°F, with an expected normal flowing pressure of 2,000 psig. A drilling mud is proposed of 349 ppb (lbm/bbl) H_2O, 124.5 ppb NaCl, 13.75 ppb bentonite, 0.63 ppb (30%) PHPA, 0.63 ppb Kelzan SC polymer (polysaccharide), 205.4 ppb barite, 0.63 ppb Drispac Super Lo, and 25 ppb simulated drill solids. Gas analyses (from neighboring wells) suggest that the gas composition is 87.2 mol% methane, 7.6 mol% ethane, 3.1 mol% propane, 0.8 mol% *n*-butane, 0.5 mol% *i*-butane, 0.2 mol% *n*-pentane, 0.2 mol% *i*-pentane, and 0.4 mol% nitrogen. Will the mud form hydrates during the drilling or if a gas kick occurs and cause a shut-in?

Solution. This is an unusual mud because the salt concentration (26.3 wt% in the water) is at saturation. Because there are no other inhibitors (e.g., salts, glycols, or glycerols) present, only the NaCl concentration in water is considered without regard for the other components. By use of HYDOFF, just as shown in Example 2.9 (but with NaCl instead of MeOH as the inhibitor), the hydrate-formation pressure at 39.8°F is predicted as 3,272 psia for the gas in this problem. The experimental value of hydrate formation in the mud is 2,895 psia at 39.8°F, indicating the predicted pressure is approximately 13% high.[44]

The mud should provide protection both at the flowing conditions (2,000 psig) and at shut-in conditions of 5,000 ft of water by use of the approximation of 2 ft/psig. In the current case, it might be good to reduce the amount of salt (with HYDOFF as a guide) to prevent salt precipitation and to provide better rheological properties.

2.10.2 Well Testing, Startup, or Restart. Davalath and Barker[3] presented a summary of dealing with hydrates in deepwater production and testing, citing two case studies of problems (Appendix C: C23 and C24) where hydrates formed in the tubing string; in one case, hydrates caused the wireline to separate upon tool retrieval. The hydrate potential is greatest when the well is cold, during startup or shut-in periods, but hydrates also can form during well flow with high pressures and low temperatures. Their report indicates three consequences of hydrate plugs during deepwater production and testing.

1. Well flow rates can be reduced or stopped.
2. Wireline tools can be blocked in the well tubing.
3. Downhole valves that control produced fluids in tubing or annulus can be blocked.

The principal tools to combat these problems are (1) adequate inhibitor injection capacity, (2) provision of two points of hydrate inhibition injection: at the subsea tree and below the seafloor, and (3) provision of insulation or heating systems to maintain high tubing temperature.

This chapter provides a method to determine hydrate formation and prevention on production testing or on the restart of a shut-in well. In well testing, there are two likely spots where hydrates might form, at the surface choke (Hydrates Caused by Expansion Through Choke Valve) and below the wellhead (Hydrates Below the Wellhead).

Hydrates Caused by Expansion Through Choke Valve. When a gas expands through a sudden restriction such as a valve, a Joule-Thomson cooling occurs. A compressible fluid is required for the cooling, so the oil, condensate, or produced water will not contribute significantly to a temperature reduction. Sec. 2.8 provides a calculation method to enable prediction of hydrate formation with such expansions and choice of a level of inhibition to prevent hydrates.

Frequently, it is possible to prevent hydrates by placing inhibited fluid just upstream of the gas expansion, as shown in the following case study.

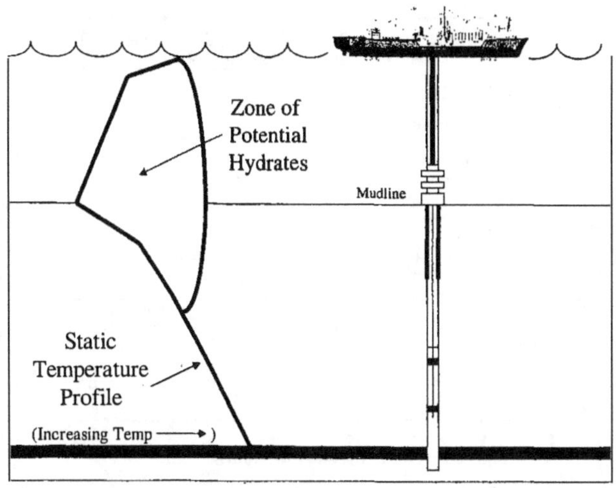

Fig. 2.48—Deepsea hydrate-formation envelope.[3]

Case Study 2.5. In Arco's Yacheng 13-1 gas field, it was possible to load brine and diesel fluid in the well differentially, not only to achieve the correct downhole pressure, but also to provide hydrate inhibition downstream of the choke.* In this field, all delineation wells experienced hydrates downstream of the choke during the initial well testing caused by underinhibited completion brines. In the delineation wells, the entire column had been loaded with fluid of the appropriate salinity (8.9 lbm/gal) to achieve the bottomhole pressure; in each case, the hydrate inhibition was insufficient to prevent hydrates downstream of the choke in the brine.

To prevent formation at the choke in development wells, they were perforated with a dual column of fluid. The lower 60% of each hole was loaded with 23% NaCl brine, while the 40% upper portion was loaded with diesel fluid to obtain the appropriate fluid density for the total column. No hydrates were experienced when the wells were perforated and brought on test, initially because of the inhibited brine (close to the cold production) and later because of increasing system temperature.

Hydrates Below the Wellhead. In addition to hydrate-plug formation by expansion through Joule-Thomson valves, a major source of hydrate problems occurs in deepwater and Arctic wells at high pressures and low temperatures below the wellhead. Particularly on startup or shutdown of well testing, the cold temperatures promote hydrate formation, for either development wells with completion brine in hole or for exploration wells with drilling fluid or produced fluids in the hole.

Fig. 2.48 shows the potential hydrate-formation envelope, defined by the intersection of the ambient temperature and pressure (or depth) and the hydrate-phase boundary.[51] In the hydrate-stability zone, the right-most limitation line (almost vertical) represents the hydrate-phase boundary. The left-most line bounding the potential hydrate zone represents the temperature and pressure profile in the well at one time during a test. Fig. 2.48 shows that the hydrate-stability envelope extends both above and below the mudline.

This section first discusses the input variables and method to calculate the potential hydrate-formation zone in the well, and then examples of the calculation are given from Exxon and Arco reports.

Input Variables and Methods in Well Testing. To prevent hydrate formation in well testing, the following variables and methods are needed.

1. The time history of the well-temperature and -pressure profiles during the test should be predicted with a transient wellbore simulation program, such as WEST or WELLTEMP, at flowing and at static conditions over the course of the test. This simulation determines the left-most boundary of the envelope shown in Fig. 2.48. The

*Personal communication with J.B. Bloys, Arco E&P Technology, Plano, Texas (11 July 1998).

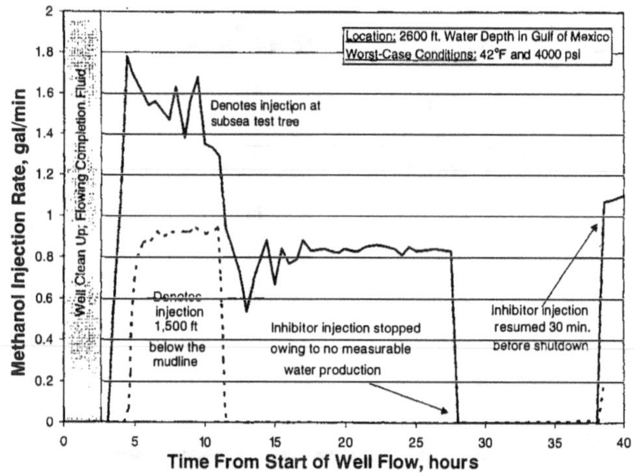

Fig. 2.49—Field Example 1: MeOH injection history.[3]

worst cases usually involve startup and shutdown, with the point most susceptible to hydrate formation being the lowest system temperature (usually at seabed temperature) and the highest (shut-in) pressure. As warm fluids are produced from the well, the temperature of the system probably is removed from the hydrate region, as indicated in Example 2.15.

2. The gas composition and the produced-water salinity should be specified. The engineer should consider gas and produced-water compositions in offset wells. If no composition data are available, assume a conservative gas composition, such as 90% methane, 7% ethane, 3% propane, and fresh water. These input data are required to find the hydrate-phase boundary, which forms the right-most line in Fig. 2.48 for the zone of hydrate stability.

3. Estimates are needed for gas, water, and liquid hydrocarbon production rate to determine the quantity of hydrate inhibitor that must be delivered. Water production consists of three sources: condensed water, produced brine, and completion and workover brine. While hydrate prevention occurs in the water phases, inhibitors such as MeOH, also are lost to the gas and hydrocarbon phases.

Input variables are usually well-defined, and the problem can be analyzed accurately for the case of a development well in a field with adequate information. In contrast, when one lacks input variables, as with an exploration well, worst-case scenarios should be used, such as the lowest seabed temperature, the highest pressure, the most propane in the gas, and fresh water to determine the concentration of inhibitor needed. Estimates of gas and water flows provide the quantity and delivery rate of inhibitor to the wellhead and subsea injection point.

Outline of General Approach to the Problem (With Data Available).

1. Plot the well temperature/pressure profile from a simulation program, such as WEST or WELLTEMP, at representative times for the test startup and shutdown.

2. List the amounts of gas, water, and condensate produced on the basis of 1 MMscf of gas. Use this figure as the basis for the calculation of the inhibitor needed. Multiply the inhibitor concentration in each phase by the number of MMscf to obtain the quantity of inhibitor in that phase.

3. By use of the gas composition and the aqueous phase salinity, determine the hydrate pressure/temperature equilibrium curves with the previous input data and the computer program HYDOFF (Appendix B.1) over the pressure range of interest for the well. See Example 2.9 for a sample calculation.

4. Use the computer program XPAND (Appendix B.2) to determine whether the expansion pressure drop across the choke valve brings the line temperature into the hydrate-formation region. Example 2.13 includes a sample calculation.

5. Use HYDOFF (as in Example 2.9) to determine the inhibitor concentration needed in the free-water phase (in addition to any brine present) to prevent hydrates either in Steps 3 or 4. The well test should be designed to prevent hydrates in both the seabed conditions and downstream of the choke.

6. Use the gas water content chart (Fig. 2.9) to determine the initial water content (at the bottomhole temperature and pressure) and the final water content of gas (at the seafloor conditions). As the gas cools from reservoir conditions, water condenses to add to the produced water. The difference between the two is the amount of water condensed from the gas-phase flow stream. See Step 2 of Example 2.10 for a sample calculation.

7. Determine the amount of inhibitor that is lost to the gas phase and to the liquid-hydrocarbon phase with Figs. 2.10 and 2.13, respectively (for MeOH), or with Figs. 2.11 and 2.14 (for ethylene glycol) as shown in Example 2.10. The concentration of inhibitor in the free-water phase (Step 5) is required in each of these figures to determine the amount of inhibitor in the gas and condensate.

8. Multiply the inhibitor concentration in Step 5 by the total water. Also multiply the inhibitor concentration in the gas and liquid hydrocarbon by their respective flow rates. The sum gives the total amount of inhibitor injection.

If the input data are uncertain, the engineer may wish to run a set of sensitivity tests to determine the effect of the range of input-data uncertainty on the amount of inhibitor. The delivery rate may be limited by the inhibitor line size. For example, the amount of ethylene glycol or MeOH delivered through a $^3/_8$-in. line may be limited by the acceptable pressure drop. The inability to deliver inhibitor may limit well-test rates. In Fig. 2.33, Davalath and Barker[3] quantify viscosity effects at various inhibitor-pump discharge pressures; MeOH or ethylene glycol are usually preferred to provide reasonable pressure drops.

As the test proceeds, the system warms and perhaps passes out of the hydrate-formation region for Joule-Thomson and normal line cooling. However, before hydrate-inhibitor rates are reduced, one should consider subsequent shut-ins and restarts during the life of the test.

Example 2.15: MeOH Injection During Production Test. In **Fig. 2.49,** Davalath and Barker[3] show the MeOH injection history for a production test of a deepwater (2,600-ft) Gulf of Mexico well. The well had the worst conditions at the seafloor at 42°F and 4,000 psi (30°F below the hydrate-formation temperature at that pressure). During the first 3 hours, no MeOH was injected because 13.4 to 15-lbm/gal completion brine was flowing. After oil and gas began to flow, MeOH injection was increased to 1.5 gal/min at the subsea tree and 1.0 gal/min at the injection point, 1,650 ft below the seafloor. Twelve hours after production was started, as tubing heating occurred, MeOH-injection rate at the tree was decreased to 0.9 gal/min and MeOH was stopped at the downhole injection point. All MeOH injection was stopped after 28 hours because the well was not producing measurable amounts of water. One-half hour before production was stopped, to inhibit the tubing as the well cooled, inhibitor injection was resumed at 1 gal/min at the subsea test tree and at 0.3 gal/min at the downhole injection point. The well was successfully tested without hydrate-formation problems.

Example 2.16: Well-Temperature Profiles During Production Test. Davalath and Barker[3] present a second example for temperature profiles during 78 hours of an exploration well startup (**Fig. 2.50**) followed by 104 hours of shut-in (**Fig. 2.51**) in 2,100 ft of water in the Gulf of Mexico. At 5,000 B/D with a 5% water cut and GOR of 800, Fig. 2.50 shows that the seafloor-tubing temperature rises from 44 to 90°F in 78 hours. During that same time, the flow is cooled by water above the seafloor, so that the surface flowline temperature drops to approximately 63°F and requires MeOH inhibition.

Fig. 2.51 shows that, on shutdown, the uninsulated tubing in the water is cooled to the original temperature (in the hydrate-formation envelope) within 2 hours (marked 80 hours), while the subsea tubing cools more slowly. On the basis of these analyses, two steps were taken to prevent hydrate formation during the test: (1) a total of 6

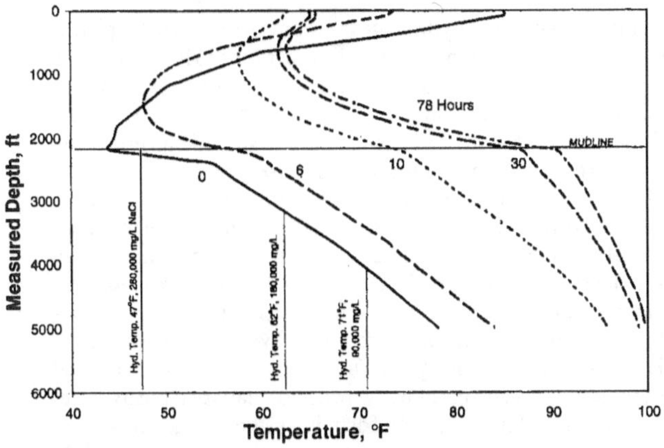

Fig. 2.50—Field Example 2: Temperature rise following production startup (uninsulated tubing).[3]

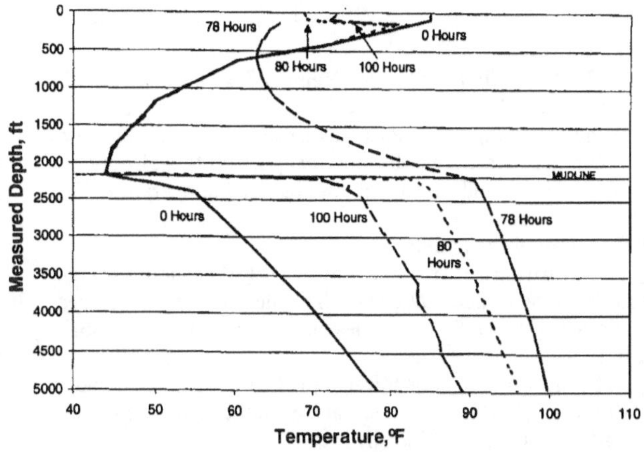

Fig. 2.51—Field Example 2: Cooldown following production shutdown (uninsulated tubing).[3]

gal/min of MeOH injection was provided to the subsea tree and to an injection point 1,500 ft below the seafloor and (2) the tubing was insulated from the surface to the seafloor to increase the surface flowline temperature by 13°F during the test. It was decided not to install insulated tubing below the seafloor because of installation difficulty for a short-term test. No hydrate problems were observed during the test.

All these principles can be applied to Arctic wells where the upper wellbore below the Christmas tree can be cold enough for hydrates to form.

2.10.3 Drilling Through Natural Hydrates in the Arctic or Seafloor.

Hydrates occur naturally in Arctic permafrost regions and in shallow ocean sediments, where plant and animal matter have been buried and decayed to produce methane (see Chap. 7 of Ref. 8). In addition, gas seeps from reservoirs have formed hydrate mounds close to the seafloor in the Gulf of Mexico and the Caspian Sea. Drilling through these hydrate zones can cause significant operational hazards as the hydrates melt and release gas. One volume of hydrate can contain as much as 170 volumes of gas at 14.7 psia and 60°F.

Hazards for drilling through hydrates are gas kicks that might blow mud out of the hole; gas release on the rig floor, with potential fire; and loosening of sediments and flow of gas around casing.

Methods used to minimize this hazard include (1) diverting all flow returns through mud degasser; (2) using the highest possible drilling mud density, limited by the fracture gradient of the formation; (3) cooling the drilling fluid to prevent hydrate dissociation; (4) increasing the mud circulation; (5) drilling with freshwater mud only; (6) drilling slowly to release the gas at a controlled rate; (7) raising the viscosity of the mud to carry the cuttings and hydrates predictably out of the hole; and (8) running the casing after penetrating the hydrate zone.

Various combinations of the first three solutions should be used on obtaining a gas kick from hydrates (e.g., closing the BOP, decreasing the mud temperature, increasing the mud density, and removing the free gas).

Goodman[52] and Goodman and Franklin[53] used computer simulation to specify fluids and to determine well-temperature and -pressure profiles while drilling through hydrates in the offshore Arctic. Their model verified that the dissociated-gas influx can be controlled by increasing mud density. Increased mud density also requires a higher hydrate-dissociation temperature by increasing the required dissociation pressure.

If the depth of the hydrate interval is known, gas influx and mud conditions can be quantified. Roadifer *et al.*[54,55] provide the nomogram given in **Fig. 2.52** to determine optimum temperatures and pressures when drilling through hydrates, either to prevent or to encourage hydrate dissociation, for an assumed gas composition. The nomogram also can be used to design surface degassing equipment if the zone of hydrate depth and the hydrate dissociation temperature is known. While this nomogram is the best currently available in the open literature, it must be validated through industrial use over time.

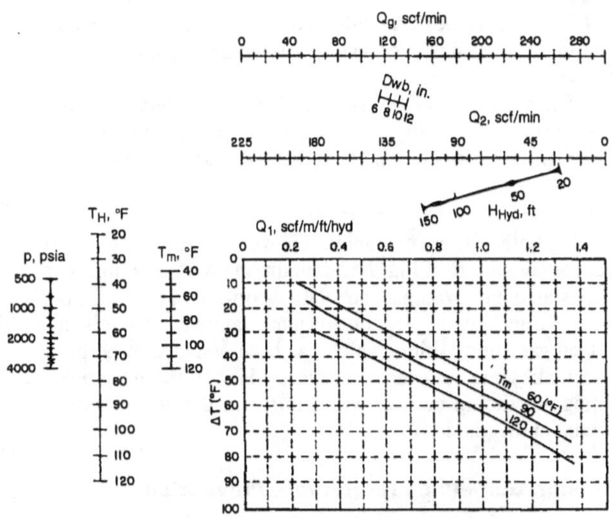

Fig. 2.52—Nomogram for hydrates drilling management.[8]

Use and definition of terms are best given by example. Two examples of the nomogram use by Roadifer *et al.*[54,55] appear next.

Example 2.17: Sizing a Mud Degasser for Drilling in Hydrate Zones. For simple hydrates of methane, determine the minimum gas capacity of the following mud degasser given hydrate-zone thickness, H_{Hyd}, 50 ft; methane hydrate-formation temperature, T_H, 60°F; wellbore diameter, D_{wb}, 8.375 in; minimum wellbore pressure, *p*, 1,750 psia; and maximum annulus temperature, T_m, 95°F.

Solution. Use of the nomogram of Fig. 2.52.

Step 1. Construct a horizontal line from the pressure, *p*, line at 1,750 psia to the hydrate-dissociation-temperature line, T_H, at 60°F.

Step 2. Construct a line from $T_H = 60$°F, through the mud-temperature line, T_m, at 95°F to the left ordinate of the crosshatched region (the ΔT axis) at $\Delta T = 35$°F.

Step 3. Construct a horizontal line from $\Delta T = 35$°F to the mud temperature within the crosshatched region, at 95°F.

Step 4. Construct a vertical line from the mud temperature to Q_1, on the top abscissa, which represents the gas released per foot of hydrate. $Q_1 = 0.55$ scf-min/ft hydrate.

Step 5. From Q_1 construct a line through H_{Hyd}, the hydrate-zone thickness (50 ft), to Q_2 the amount of gas released. Read $Q_2 = 28$ scf/min.

Step 6. Construct a line from Q_2 through the diameter of the wellbore D_{wb} (8.375 in.) to Q_g to account for restrictions in the wellbore. Read the maximum gas rate (minimum capacity of degasser required) at 28 scf/min.

Example 2.18: Wellbore Pressure Required To Limit Dissociation. Find the wellbore pressure required to limit hydrate dissociation for the following conditions: gas capacity of mud degasser, Q_g, 75 scf/min; hydrate zone thickness, H_{Hyd}, 100 ft; wellbore diameter, D_{wb}, 11 in.; and maximum annulus-mud temperature, T_m, 100°F.

Solution. Use of the nomogram of Fig. 2.52.

Step 1. Construct a line from the top line Q_g (80 scf/min) through the diameter of the wellbore D_{wb} (11 in.) to intersect the gas released from hydrate Q_2 at 75 scf/min. Q_2 lies on the second top-most horizontal line.

Step 2. From Q_2 draw a line through the hydrate-zone thickness, H_{Hyd} (100 ft) to Q_1, the top horizontal line in the crosshatched region. Read Q_1 at 0.74 scf-min/ft hydrate.

Step 3. Draw a vertical line from Q_1 to the mud temperature ($T_m = 100$°F) in the crosshatched region. Draw a horizontal line from the mud temperature to the left abscissa ($\Delta T = 44$°F) on the crosshatched region.

Step 4. From the value of $\Delta T = 44$°F, draw a line through the mud temperature (100°F on the vertical T_m line) to the vertical hydrate-temperature line T_H. Read the intersection at 56°F.

Step 5. Draw a horizontal line from the hydrate temperature to the left-most vertical line, p, the minimum wellbore pressure required to prevent hydrate dissociation. Read that value as 1,400 psia.

In closing both of these examples, it is worthwhile to re-emphasize that the nomogram of Fig. 2.52 contains many assumptions not explicitly stated. For example, the relative position of the left-most two lines relating hydrate pressure and temperature are strongly dependent on gas composition, which has been assumed to be pure methane. Until some track record of success is obtained in the field, the nomogram should be used with caution as the best literature approximation available.

2.11 Design Guidelines for Hydrate Prevention

The following hydrate-prevention paradigm is a collection of heuristics that provides general guidelines for design. These design guidelines are for hydrate prevention in the three parts of the system where hydrates most often occur—the well, the pipeline, and the process or platform. Many of these guidelines result from Chap. 3 on Hydrate Remediation.

1. Obtain the systems' hydrate-formation curve. Before embarking on a hydrate-prevention design, it is imperative to have a reliable hydrate-equilibrium curve (Secs. 2.6 and 2.7) that represents the gas, oil/condensate, and water compositions over the life of the field. If possible, the hydrate-formation curve should be verified by means of several prediction programs or experiments.

2. Simulate the pressure/temperature profile in the well, flowlines, and platform at the worst case (usually during winter months) over the life of the field. Estimate the water-residence times at all points in the system. Account for both normal cooling (e.g., in pipelines, as in Example 2.2) and Joule-Thomson expansions across restrictions (e.g., in wells, chokes, and control valves, as in Sec. 2.8).

3. Estimate the subcooling ΔT (at the lowest temperature and highest pressure) at each point in the process relative to the hydrate-equilibrium curve. Hydrates may form in systems with subcooling ΔT's less than 2 to 4°F.

4. Where subcooling is unavoidable, determine the type of hydrate inhibition, such as chemical-inhibitor injection (Secs. 2.9.1 through 2.9.3, particularly Guidelines for Use of Kinetic of Thermodynamic Inhibitors) or heat management (Sec. 2.9.4). Choose the inhibition method with regard to both prediction ability and operating experiences. Consider providing a heater before the platform choke and separator.

5. Eliminate subcooling points of likely hydrate formation. Design pipelines to minimize buckling and protrusions from the earth that might promote cooling.

6. Design large pressure drops with either dry gas or expansions at high-temperature points in the process. Where large expansions of wet gas are unavoidable (e.g., at choke valves), provide MeOH injection capability upstream of the restriction.

7. Eliminate points of water accumulation, such as upslopes in pipelines or "S" configurations in risers. Where pipeline topography ensures water accumulations (e.g., upstream of upslopes in lines), provide for frequent pigging and consider placing MeOH injection before the accumulation points.

8. Eliminate points of hydrate accumulation from a mechanical perspective. Hydrate crystals in a line accumulate (and plug) wherever light sand particles might accumulate, such as at blind flanges, elbows, screens and filters, upstream of restrictions, and other such areas. Avoid unnecessary bends. Bend radii less than five times the pipe diameter should be avoided to facilitate coiled-tubing entry. An offshore platform-riser tube radius should be from 20 to 80 ft.

9. Design water-based drilling fluids recognizing that the water and the water-soluble drilling-fluid components determine the conditions of hydrate formation. Concentration of salts, alcohols,

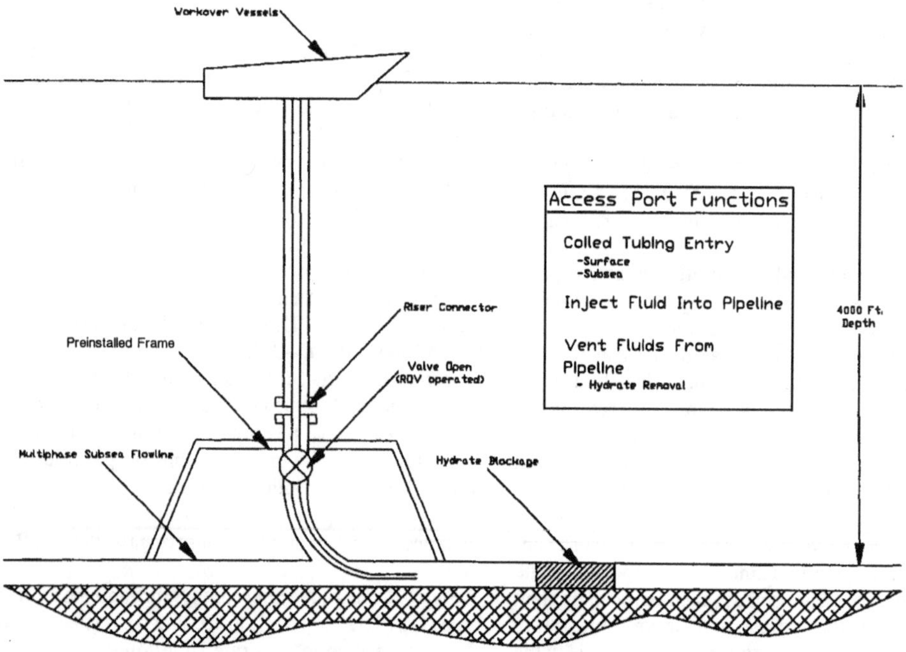

Fig. 2.53—Preinstalled access ports on pipeline (ROV=remotely operated vehicle).[56]

glycols, and glycerols in the water determine the hydrate-formation temperature and pressure.

10. Plan a contingency for hydrate formation. A high probability exists that hydrates will form over the system lifetime. Provide hydrate-remediation methods (see Chap. 3 for justification) in the design.

a. For pipelines, safe remediation often implies depressurization from both ends of a hydrate plug. Optimally, multiple access points in a pipeline (see b) are invaluable in locating and remediating hydrate plugs. Alternatively, dual production lines should be used to provide for depressurization of wellheads from the upstream side of a pipeline plug.

As a second-best method, provide for depressurization through a wellhead service line (for corrosion, paraffin, or hydrate inhibitor injection) with bypass capability for check valve(s) at the point of injection. As a minimum, a spare flange and valve should be provided at the wellhead or manifold so that depressurization can be done by connection to offshore production vessel (see Arco Case Study 3.6 in Sec. 3.4.1). Technology is not yet available for accurate location of the end of plugs or the safe heating of plugs in ocean pipelines.

b. Subsea access points should be considered at the well manifold and at 4-mile intervals along the pipeline (**Fig. 2.53**). Such access points facilitate (1) the location of a hydrate, (2) venting of excessive fluid head from plugs in deepsea lines, (3) injection of hydrate inhibitors, (4) coiled-tubing entry, and (5) pig launching.

c. In wells, hydrate remediation occurs by approaching one end of a plug by means of chemical injection, depressurization, heating, or coiled tubing. Case Study 3.7 in Sec. 3.4.3 indicates that normal well lubricators can be used at the swab valve with careful balancing of pressure.

d. Above ground (or above water) hydrate plugs may be located with tools, such as a thermocamera (see Fig. 3.7 of Sec. 3.3.1). With the accurate location of the plug ends, remediation may be done through chemical injection, heating, or depressurization.

Nomenclature

C_I = constant for a particular inhibitor
D_{wb} = wellbore diameter, L
H_{Hyd} = hydrate-zone thickness, L
ΔH = enthalpy change, mL^2/t^2
k_I = kinetic constant for a particular inhibitor
K^L_{MEG} = MEG mole fraction in condensate relative to water
K^L_{MeOH} = MeOH mole fraction in condensate relative to water
K^V_{MeOH} = ratio of MeOH mole fraction in gas relative to water
L = length, L
L_W = liquid water phase
m_{MeoH} = mass of MeOH
m_W = mass of water
M = molecular weight
M_I = molecular weight of inhibitor I
$M_{MeOH\%}$ = molecular weight percent of MeOH
M_{MeOH} = molecular weight for MeOH
M_{MEG} = molecular weight for MEG
M_a = air molecular weight
M_g = gas molecular weight
p = minimum wellbore pressure, m/Lt^2, psi
p_{eq} = hydrate equilibrium pressure, m/Lt^2, psi
p_{in} = inlet pressure, m/Lt^2, psi
p_{out} = outlet pressure, m/Lt^2, psi
Q = heat added, mL^2/t^2
Q_g = gas capacity of mud degasser, L^3
Q_1 = gas released per ft of hydrate, L^3/L
Q_2 = amount of gas released, L^3
T = inhibitor temperature depression, T, °F
T_{eq} = hydrate-equilibrium temperature, T, °F
T_H = methane hydrate-formation temperature, T, °F
T_m = mud temperature line, T, °F
T_{op} = operating temperature, T, °F
T_2 = outlet temperature, first expansion valve, T, °F
T_3 = outlet temperature, second expansion valve, T, °F
T_4 = outlet temperature, third expansion valve, T, °F
ΔT = hydrate depression temperature, T, °F
ΔT_{max} = maximum value of subcooling, T, °F
U = heat-transfer coefficient, m/t^3T
w = weight in pounds
w_{MeOH} = weight of MeOH in pounds
W_I = weight percent inhibitor
$W_{I,L}$ = weight percent of the inhibitor in the liquid
W_W = weight of water
W_s = shaft work, mL^2/t^2
x_{MEG} = mole fraction of MEG in water phase
x_{MeOH} = mole fraction of MeOH in water phase
y_i = mole fraction of i in the vapor phase
y_{MEG} = mole fraction of MEG in vapor phase
y_{MeOH} = mole fraction of MeOH in vapor phase
z_{MEG} = mole fraction of MEG in condensate
z_{MeOH} = mole fraction of MeOH in condensate
σ = liquid-crystal surface tension, m/t^2
γ_g = gas gravity

References

1. Churgin, J. and Halminski, S.J.: "Waters Off the United States, Vol. II, Gulf of Mexico," *Key To Oceanographic Records Documentation #2*, Natl. Oceanographic Data Center, Washington D.C., 1974.
2. Davalath, J. and Barker, J.W.: "Hydrate Inhibitor Design for Deepwater Completions," paper SPE 26532 presented at the 1993 SPE Annual Technical Conference and Exhibition, Houston, 3–6 October.
3. Barker, J.W.: "Formation of Hydates in Deepwater Drilling Operations," paper presented at N.L. Baoid Meeting on Hydrates in Deepwater Drilling Muds, Houston, 20 October 1988.
4. Barker, J.W. and Gomez, R.K.: "Formation of Hydrates During Deepwater Drilling Operations," *JPT* (March 1989) 297.
5. Notz, P.K.: *Technical Memorandum TM 94-0145 on Relative Inhibitor Costs*, Texaco (26 July 1994).
6. Manning, F.S. and Thompson, R.E.: *Oilfield Processing of Petroleum: Volume One: Natural Gas*, Pennwell Publishing, Tulsa, Oklahoma (1991) 80–82, 344–355.
7. Todd, J.L. *et al.*: *Reliability Engineering—Gas Freezing and Hydrates*, Texaco (1996).
8. Sloan, E.D. Jr.: *Clathrate Hydrates of Natural Gases*, second edition, Marcel Dekker Inc., New York (1998).
9. Dendy, J.W.: "Trouble Shooting and Preventing Maintenance of Offshore Gas Systems," *Proc.*, Fifth Annual Gas Proc. Assn., Lafayette, Louisiana (1969) 161.
10. Katz, D.L. *et al.*: *Handbook of Natural Gas Engineering*, McGraw-Hill Book Co. Inc., New York (1959).
11. Katz, D.L.: "Prediction of Conditions for Hydrate Formation in Natural Gases," *Trans.*, AIME (1945) **160**, 140.
12. Sloan, E.D. Jr.: "The Colorado School of Mines Hydrate Program," *Proc.*, 64th Annual GPA Convention, Gas Processors Assn., Houston (March 1985).
13. Hammerschmidt, E.G.: "Gas Hydrate Formations in Natural Gas Pipelines," *Oil & Gas J.* (1939) **37**, No. 50, 66.
14. Nielsen, R.B. and Bucklin, R.W.: "Use of Methanol for Hydrate Control in Expander Plants," *Proc.*, Gas Conditioning Conference, Norman, Oklahoma (1983).
15. McKetta, J.J. and Wehe, A.H.: "Use This Chart for Water Content of Natural Gases," *Petroleum Refiner* (1958) **37**, 153.
16. Townsend, F.M. and Reid, L.S.: *Hydrate Control in Natural Gas Systems*, L.S. Reid and Assocs., Norman, Oklahoma (23 September 1978).
17. Robinson, D.B. and Ng, H.-J.: "Hydrate Formation and Inhibition in Gas or Gas Condensate Streams," *J. Cdn. Pet. Tech.* (1986) **25**, No. 4, 26.
18. Baillie, C. and Wichert, E.: "Chart Gives Hydrate Formation Temperature for Natural Gas," *Oil & Gas J.* (April 1987) 37.
19. Ng, H.-J. and Robinson, D.B.: "Equilibrium Phase Composition and Hydrating Conditions in Systems Containing Methanol, Light Hydrocarbons, Carbon Dioxide, and Hydrogen Sulfide," GPA Research Report RR-66, Gas Processors Assn. (joint with Canadian Gas Processors Assn.) Tulsa, Oklahoma (April 1983).
20. Ng, H.-J. and Chen, C.-J.: "Vapour-Liquid and Vapour-Liquid-Liquid Equilibria for H_2S, CO_2, Selected Light Hydrocarbons, and a Gas Condensate in Aqueous Methanol or Ethylene Glycol Solutions," GPA Re-

search Report RR-149, Gas Processors Assn., Tulsa, Gas Research Inst., Chicago (1995).

21. Polderman, L.D.: "The Glycols as Hydrate Point Depressant in Natural Gas Systems," *Proc.*, Gas Conditioning Conference, Norman, Oklahoma (1958).
22. Brown, G.G.: "A Series of Enthalpy-Entropy Charts for Natural Gases," *Trans.*, AIME (1945) **160,** 65.
23. Loh, J., Maddox, R.N., and Erbar, J.H.: "New Hydrate Formation Data Reveal Differences," *Oil & Gas J.* (1983) **81,** No. 20, 96.
24. Deaton, W.M. and Frost, E.M. Jr.: *Gas Hydrates and Their Relation to the Operation of Natural-Gas Pipe Lines*, Monograph 8, U.S. Bureau of Mines (1946).
25. Yousif, M.H. *et al.*: "The Effects of Low Concentration Methanol Solutions on Hydrate Formation," *Proc.*, Second Intl. Conf. on Natural Gas Hydrates, Toulouse, France (2–6 June 1996) 291.
26. Makogon, Y.F.: *Hydrates of Natural Gas*, Pennwell Publishing, Tulsa, Oklahoma (1981) 133.
27. Stange, E., Majeed, A.I., and Overa, S.: "Experimentations and Modeling of the Multiphase Equilibrium and Inhibition of Hydrates," *Proc.*, 68th Annual GPA Convention, San Antonio, Texas (March 1989).
28. "Evaluation of the Impact of Methanol Recycle/Recovery Facilities on Offshore Production Systems," Report DeepStar II CTR 221-1, Paragon Engineering Services Inc. (December 1994).
29. Ng, H.-J., Chen, C.-J., and Robinson, D.B.: "The Influence of High Concentrations of Methanol on Hydrate Formation and the Distribution of Glycol in Liquid-Liquid Mixtures," GPA RR-106, Gas Processors Assn., Tulsa, Oklahoma (April 1987).
30. Lingelem, M.N., Majeed, A.I., and Stange, E.: Industrial Experience in Evaluation of Hydrate Formation, Inhibition, and Dissociation in Pipeline Design and Operation," E.D. Sloan Jr., J. Happel, and M. Hnatow (eds.), First Intl. Conf. on Natural Gas Hydrates, New York Academy of Sciences, (1994) **715,** 75.
31. Behar, E. *et al.*: "Plugging Control of Production Facilities by Hydrates," First Intl. Conference on Natural Gas Hydrates, E.D. Sloan, J. Happel, and M.A. Hnatow (eds.) New York Academy of Sciences, (1994) **715,** 94.
32. Bloys, J.B., Lacey, C., and Lynch, P.: "Laboratory Testing and Field Trial of a New Kinetic Hydrate Inhibitor," paper OTC 7772 presented at the 1995 Offshore Technology Conference, Houston, 1–4 May.
33. Notz, P.K. *et al.*: "The Application of Kinetic Inhibitors to Gas Hydrate Problems," *SPEPF* (November 1996) 256.
34. Argo, C.B. *et al.*: "Commercial Deployment of Low-Dosage Hydrate Inhibitors in a Southern North Sea 69-km Wet-Gas Subsea Pipeline," paper SPE 37255 presented at the 1997 SPE Intl. Symposium on Oilfield Chemistry, Houston, 18–21 February.
35. Corrigan, A. *et al.*: "Trials of Threshold Hydrate Inhibitors in the Ravenspurn to Cleeton Line," *SPEPF* (November 1996) 250.
36. "Insulated Flowline Options Identification Study," Report DeepStar IIA CTR A601-a, INTEC Engineering Co., (August 1995).
37. "Deepwater Gulf of Mexico Integrated Towed Flowline Bundle Design Study," Report DeepStar IIA CTR A601-b, Kvaerner—R.J. Brown, (December 1995).
38. "Deepwater Gulf of Mexico Insulated Pipe-in-Pipe Study," Report DeepStar IIA CTR A601-c, Snamprogetti, SpA Offshore Div., (December 1995).
39. Davalath, J.: "Non-Jacketed Insulation Coating," Report DeepStar IIA CTR A601-d, Mentor Subsea (February 1996).
40. Barkley, R.W., Dean, T.L., and Kashou S.:"Deepstar Insulation Requirement for Pipeline Bundles," Report DeepStar II CTR 223-1, Texaco Central Engineering and Purchasing Dept. and Texaco E&P Technology Dept. (June 1994).
41. Aarseth, F.: "Use of Electrical Power in Control of Wax and Hydrates," paper OTC 8541 presented at the 1997 Offshore Technology Conference, Houston, 5–8 May.
42. Grigg, R.B. and Lynes, G.L.: "Oil-Based Drilling Mud as a Gas Hydrate Inhibitor," *SPEDE* (March 1992) 32.
43. Hale, A.H.: "Shell's Strategy for Gas Hydrate Inhibition," paper presented at N.L. Baroid 1988 Hydrates in Deep Water Drilling Meeting, Houston, 20 October.
44. Kotkoskie, T.S. *et al.*: "Inhibition of Gas Hydrates in Water-Based Drilling Muds," *SPEDE* (June 1992) 130.
45. Ouar, H. *et al.*: "The Formation of Natural Gas Hydrates in Water-Based Drilling Fluids," *Trans.*, I. Chem. E. (January 1992) **70(A)** 48.
46. Cha, S.B. *et al.*: "A Third Surface Effect on Hydrate Formation," *J. Phys. Chem.* (1988) **92,** 6492.
47. Skalle, P. and Podio, A.L.: "Trends Extracted from 1,200 Gulf Coast Blowouts During 1960–1996," *World Oil* (June 1998) 67.
48. Hale, A.H. and Dewan, A.K.R.: "Inhibition of Gas Hydrates in Deepwater Drilling," *SPEDE* (June 1990) 109.
49. Lai, D. and Dzialowski, A.K.: "Investigation of Natural Gas Hydrates in Various Driling Fluids," paper SPE 18637 presented at the 1989 SPE/IADC Drilling Conference, New Orleans, February–March.
50. Ebeltoft, H., Yousif, M.H., and Soergaard, E.: "Hydrate Control During Deepwater Drilling: Overview and New Drilling Fluids Formulations," paper SPE 38567 presented at the 1997 SPE Annual Technical Conference and Exhibition, San Antonio, Texas, 6–8 October.
51. Walz, G.S.: "Monoethylene Glycol Injection Worksheet, Ver. 1.2," internal report, Arco E&P Technology, Plano, Texas (March 1998).
52. Goodman, M.A.: "Handbook of Arctic Well Completions," *World Oil* (1978) 13.
53. Goodman, MA. and Franklin, L.J.: *Proc.*, Fourth Canadian Permafrost Conference, H.M. French (ed.), Cdn. Natl. Research Council (1982) 349.
54. Roadifer, R.D., Godbole, S.P., and Kamath, V.A.: "Thermal Model for Establishing Guidelines for Drilling in the Arctic in the Presence of Hydrates," paper SPE 16361 presented at the 1987 SPE California Regional Meeting, Ventura, California, 8–10 April.
55. Roadifer, R.D., Godbole, S.P., and Kamath, V.A.: "Estimation of Parameters for Drilling in the Arctic and Offshore Environments in the Presence of Hydrates," paper SPE 16671 presented at the 1987 SPE Annual Technical Conference and Exhibition, Dallas, 27–30 September.
56. Davalath, J.: "Methods To Clear Blocked Flowlines," Report DeepStar IIA CTR A208-1, Mentor Subsea (January 1996).

SI Metric Conversion Factors

bbl	× 1.589 873	E − 01	= m^3
°F	(°F–32)/1.8		= °C
ft	× 3.048*	E − 01	= m
ft^3	× 2.831 685	E − 02	= m^3
gal	× 3.785 412	E − 03	= m^3
in.	× 2.54*	E + 00	= cm
lbm	× 4.535 924	E − 01	= kg
mile	× 1.609 344*	E + 00	= km
psi	× 6.894 757	E + 00	= kPa

*Conversion factor is exact.

Chapter 3
Hydrate-Plug Remediation

3.1 Introduction

Perhaps the best way to learn how to remove a hydrate blockage in a flow channel is to use the experiences of those who have removed similar blockages. In addition to the case studies cited in the body of the book, Appendix C details 27 case histories of hydrate removal in flowlines. **Table 3.1** provides an overview of Appendix C case histories.

Rule-of-Thumb 15. Hydrate blockages occur owing to abnormal operating conditions such as well tests with water, loss of inhibitor injection, dehydrator malfunction, startup, and shut-in. In all recorded instances,* pipeline plugs caused solely by hydrates were removed when proper safety precautions were observed.

Chap. 1 details the severe safety concerns that accompany Rule-of-Thumb 15. Yet, with proper safety precautions, no pipelines were abandoned or replaced because of a hydrate plug, as is sometimes the case for offshore paraffin plugs. Because every hydrate plug is unique, however, individual case studies are anecdotal in nature. A very large number of anecdotal studies are required before detailed remediation rules of thumb can be stated with confidence.

Fortunately, three systematic studies of hydrate plugs provide substantial guidelines for remediation.

1. In 1994, Statoil purposely formed/removed more than 20 hydrate plugs in a 6-in. gas/condensate line during a 9-week period (Case Studies C.15 through C.17 in Appendix C). During 1995–96, uninhibited plug formations were studied as baselines for new inhibitors.
2. Mentor Subsea compiled DeepStar IIA Report A208-1 "Methods To Clear Blocked Flowlines[1]" to document 16 hydrate-blockage cases and 39 paraffin-blockage cases.
3. In February 1997, DeepStar sponsored a study to form and dissociate hydrates in a Kerr-McGee field line, resulting in tests (Case Studies C.25 through C.27) with extensive instrumentation at five pipeline points.[2]

Much of the information in this chapter on hydrate remediation was excerpted from these three systematic studies, supplemented by the literature and personal interviews relating to hydrate blockages. The sections are organized to provide answers to the following questions:

3.2 How Do Hydrate Blockages Occur?

3.3 How Can Hydrates Be Detected and Located?

3.4 How Can Hydrate Plugs Be Removed?

3.5 How Can Hydrates Be Prevented on Flowline Shut-In/Startup?

*An exception was the LASMO Staffa subsea field in the North Sea, which was abandoned in 1995 because of low production problems resulting from combined waxes and hydrates. See Case Study C.6. for further history of this field, which included a 1-mile-flowline replacement.

3.2 How Do Hydrate Blockages Occur?

Fig. 2.1 in Chap. 2 shows a simplified offshore process between the well inlet and the platform-export discharge. Fig. 2.7 and accompanying discussion describe the onshore process points of hydrate formation. Sec. 2.4 illustrates hydrate-prevention design where virtually all hydrates occur—namely in the well, the pipeline, or the process/platform. In normal conditions, high reservoir temperatures before the well prevent hydrates; after the process or platform, transportation lines are dry, with insufficient water to form hydrates.

The system temperature and pressure at the point of hydrate formation must be within the hydrate-stability region, as determined by the methods in Secs. 2.6 and 2.7. For hydrates to form, the system temperature and pressure must first enter the hydrate-formation region, either through a normal cooling process (Example 2 and Figs. 2.4 and 2.5) or through a Joule-Thomson process (Sec. 2.8).

The rate of hydrate formation is a function of the degree of subcooling, ΔT (see Fig. 2.38 in Sec. 2.9.2), relative to the hydrate-formation line. Hydrates can form with subcooling ΔT less than 2 to 4°F, particularly in industrial systems where contaminants like sand, weld slag, etc., are present to serve as nucleation centers. However, hydrates with such a low degree of subcooling form more slowly than in systems that have a subcooling of 10°F or more.

3.2.1 Concept of Hydrate-Particle and -Blockage Formation. All natural-gas hydrates are approximately 85 mol% water and 15 mol% natural gas. Hydrate formation always occurs at the hydrocarbon/water interface, because this 85 : 15 ratio is far in excess of the solubility of gas in the bulk water (< 0.06 mol%) or water in the bulk gas (< 4%). This exceptionally low mutual solubility is the result of the polar nature of water.[3]

When hydrate particles occur in a static system, a solid hydrate shell forms an impenetrable barrier at the hydrocarbon/water interface, which prevents further contact of the hydrocarbon and water phases. Diffusion through the solid is extremely slow, and hydrate fissures or cracks provide the most rapid means for further contact of the water and hydrocarbon. Because of the hydrate-formation barrier at the interface, natural-gas-hydrate particles have water as an occluded phase, so that small water droplets or balloons are created within hydrate shells. Infrequently, when gas is bubbled through water, gas is the occluded phase within a hydrate shell.

TABLE 3.1—SUMMARY OF HYDRATE-BLOCKAGE EXPERIENCES IN APPENDIX C

Case/ Operator	Field/ Region	Line Size (in.)	Line Type	Well Depth (ft)	Time
1. Placid	Green Canyon 29	16	Gas condensate	200	1989
2. Chevron	Wyoming	4	Gas condensate	0	Winter
3. Chevron	Gulf of Mexico	—	Gas lift injection lines	0	Winter
4. Chevron	Oklahoma	4	Gas sales	0	1995
5. Chevron	Canada	6	Gas condensate export	0	Winter
6. Lasmos	North Sea	8	Multiphase	?	1994
7. Texaco	Garden Banks 189	2¾	Gas	725	1995
8. Texaco	Green Canyon 6	¾	Gas	600	1992
9. Texaco	North Sea	¼	Instrument valve	—	—
10. Elf Norge	N.E. Frigg	16	Gas condensate	—	1990
11. Marathon	EB 873	—	Gas export	800	1995
12. Phillips	Cod North Sea	16	Gas and condensate export	—	1978
13. Texaco	Wyoming	—	Gas	0	1995
14. Texaco	East Texas	4 to 6	Gas	0	1995
15. Statoil	Tommeliten North Sea	6	Gas condensate	—	1994
16. Statoil	Tommeliten	6	Gas condensate	7.15 miles	1994
17. Statoil	Tommeliten	6	Gas condensate	7.15 miles	1994
18. Oxy	North Sea	—	Gas condensate	—	—
19. Amoco	North Sea	—	Gas export	70 miles	—
20. Petrobrás	Brazil	—	Manifold	0	—
21. Exxon	California	—	Well	0	1989
22. Exxon	Gulf of Mexico	—	Well	0	1989
23. Exxon	South America	—	Well	0	1993
24. Exxon	Gulf of Mexico	—	Well abandonment	0	1993
25. Kerr-McGee	Wyoming	—	Gas condensate	3 mile	1997
26. Kerr-McGee	Wyoming	—	Gas condensate	3 mile	1997
27. Kerr-McGee	Wyoming	—	Gas condensate	3 mile	1997

In a turbulent system, such as a pipeline, high agitation rates provide for surface renewal, which can cause hydrate particles and agglomerations to build up and obstruct pipe flow. Over time, the particle agglomeration consolidates and loses permeability and porosity.[4]

Rule-of-Thumb 16. In gas/water systems, hydrates tend to form on the pipe wall. In gas/condensate or gas/oil systems, hydrates frequently form from free water as particles that agglomerate and bridge as larger masses in the bulk streams.

Rule-of-Thumb 16 was obtained through multiple studies on flow loops/wheels at Statoil's Research Center in Trondheim, Norway. In gas systems, water may splash or adsorb on the pipe wall where hydrate nucleation and growth may occur. In an oil/condensate system, the light-hydrocarbon liquid above the water prevents splashing and causes hydrate-particle formation and agglomeration at the liquid/liquid interface.

In a black-oil system, often only a small amount (less than 5 vol%) of water forms hydrates, but all the water and condensate are trapped in the open, porous system and can form a blockage.* In Statoil's Tommeliten field, blockages formed from a hydrate slurry with <1 vol% of the water present. Such results are fluid-dependent; while some oil/water systems convert to hydrates almost immediately with fairly low water conversion, other oil systems are more difficult to convert, but practically all water might be transformed to hydrates.

Rule-of-Thumb 17. Agglomeration of individual hydrate particles causes an open hydrate mass that has a high porosity (typically $>50\%$) and is permeable to gas flow (permeability/length ratio of $8.7:11\times10^{-15}$ m). Such an open hydrate mass has the unusual property of transmitting pressure while being a substantial liquid-flow impediment. Hydrate particles anneal to lower permeability over longer times.

*Personal communication with O. Urdahl, Statoil Research Center, Trondheim, Norway (14 July 1997).

Rule-of-Thumb 17 was obtained through both field and laboratory studies at Statoil's Tommeliten Gamma field and SINTEF's research center.[4,5] Plug porosity is determined by forming conditions and fluid effects; some plugs can have porosities considerably higher than 50%, while other plug porosities can be considerably lower. Because liquid surface tension is higher than that of gas by a factor of approximately 1,000, hydrate plugs are much less permeable to liquid than to gas.

Fig. 3.1 is a schematic of the case of hydrate formation along the wall periphery in a gas system. This slow buildup of hydrates along the wall may be characterized by the gradual increase in line Δp witnessed in two of three DeepStar field tests in a Wyoming gas-condensate line.[2]

Fig. 3.2 shows the case of hydrate formation as agglomerating or bridging particles in a condensate or oil system, providing the open, porous structure. The Statoil experience suggests that Fig. 3.2 represents a very common case of hydrate formation. However, two schools of thought exist about hydrate formation: (1) the gradual buildup of hydrate formation on the walls results in the less-porous plugs seen in a few thoroughly instrumented DeepStar field tests (See Case Studies C.25 through C.27 with gas and condensate) and (2) the multitude of Statoil studies that suggests a high-porosity, bridging hydrate-particle structure may be the norm (See Case Studies C.15 through C.17 with gas and black oil).

The number of state-of-the-art hydrate studies in field pipelines is too small to determine the causes and frequency of either type of hydrate buildup. It is apparent from the small number of studies, however, that a wide range of hydrate porosities may be attained.

The porosity/permeability of hydrate plugs largely determines their remediation. For example, if a hydrate plug is depressurized from only one end, gas flow through the plug causes Joule-Thomson cooling, as in Example 2.12, so that the downstream side of the plug may be in the hydrate-formation region at the lower temperature. This effect has been observed at the Tommeliten field[5] and provides a technical reason (added to the safety reasons in Chap. 1) for depressuring a plug from both sides. However, Case Studies C.25

TABLE 3.1—SUMMARY OF HYDRATE-BLOCKAGE EXPERIENCES IN APPENDIX C (continued)

Case/ Operator	Extent of Restriction	Control Method Before Plug	Operations Before Plug	How Was Plug Removed?	Current Prevention Method
1. Placid	Complete	None	Flowing	Depressurize	Gas dehydration Inject MeOH
2. Chevron	Complete	Heating tape and insulation	Flowing	Depressurize Heat MeOH	Inject MeOH
3. Chevron	Complete	None	Flowing	Inject MeOH	Inject MeOH Vary flow rate
4. Chevron	Partial	None	—	Depressurize Inject MeOH Heat gas	Remove restriction at flowmeter MeOH during winter
5. Chevron	Complete	None	Shut in for several days	Depressurize Heat with welding rig	Depressurize after 24 hours shut-in
6. Lasmos	Complete	MeOH	Flowing	Replace 1.24-mile blocked section	Same as before Planning to abandon
7. Texaco	Complete	None	Flowing	Depressurize Inject MeOH	Inject MeOH Gas dehydration
8. Texaco	Partial	None	Flowing	Depressurize Inject MeOH	Inject MeOH Gas dehydration
9. Texaco	Complete	None	Flowing	Inject MeOH	Occasionally inject MeOH
10. Elf Norge	Partial	MeOH	Flowing	Depressurize Inject MeOH	Inject MeOH
11. Marathon	Partial	Inadequate MeOH	Flowing	Inject more MeOH	Maintain Inject MeOH
12. Philips	Complete	MeOH pigging	Pig stuck	Depressurize Inject MeOH	Inject MeOH Dehydrate condensate
13. Texaco	Complete	MeOH	Field test	Depressurize	Combination of KI and MeOH
14. Texaco	Complete	MeOH	Field test	Depressurize	Combination of KI and MeOH
15. Statoil	Complete	MeOH injection	Flowing/ shut in/ restart	Depressurize Inject MeOH	Continue MeOH injection
16. Statoil	Complete	MeOH	Field study	Depressurize Inject MeOH	—
17. Statoil	Complete	None	Field study	Depressurize	—
18. Oxy	Complete	MeOH	Flowing	Depressurize Inject MeOH	Ensure proper MeOH injection
19. Amoco	Complete	None (Dry)	Flowing	Depressurize Inject MeOH	Ensure dehydration
20. Petrobrás	Complete	Ethanol	Startup	Depressurize Inject Ethanol	Drain manifold of water before startup
21. Exxon	Complete	None	Drilling	—	—
22. Exxon	Complete	None	Drilling	—	—
23. Exxon	Complete	None	Testing	Coiled tubing Hot glycol	—
24. Exxon	Complete	Methanol	Shut-in	Hot $CaBr_2$	—
25. Kerr-McGee	Complete	Methanol	Intensional formation	Depressurize	Methanol
26. Kerr-McGee	Complete	Methanol	Intensional formation	Depressurize	Methanol
27. Kerr-McGee	Complete	Methanol	Intensional formation	Depressurize	Methanol

through C.27 detail safe techniques for depressuring one side of a hydrate plug in DeepStar Wyoming field studies.[2]

Fig. 3.3 shows two types of pressure-drop, Δp, increases that occur with hydrate blockage of lines. On the left, Fig. 3.3a shows the gradual increase in Δp that would occur if hydrates formed on walls, giving an ever-decreasing annulus as shown in Fig. 3.1. Fig. 3.3b shows the more typical case of multiple spikes in Δp before the final plug forms; these spikes indicate that small hydrate particles form blockages, then release, as depicted by particle agglomeration in Fig. 3.2.

3.2.2 Points of Hydrate Blockage. The previous conceptual picture of hydrate formation reinforces field experiences regarding points in processes where hydrate formation occurs. For example, subcooling occurs where pipelines protrude from the earth or from mudlines, so dips in pipelines should be minimized. Large pressure drops (e.g., at orifices/valves) should be avoided.

Points of water accumulation, such as "S" configurations in pipelines or risers, also should be minimized. Where pipeline topography ensures water accumulations (e.g., upslopes in lines), providing pig and inhibitor injection points may be considered as preventative measures. Hydrate particles in a line may accumulate (and plug) wherever light sand particles might accumulate, such as at blind flanges, elbows, short-radius bends, screens and filters, and upstream of restrictions.

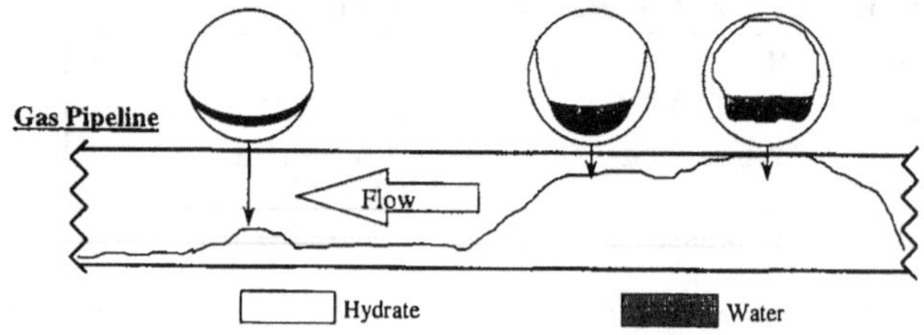

Fig. 3.1—Hydrate accumulation in gas pipeline (Lingelem *et al.*[6]).

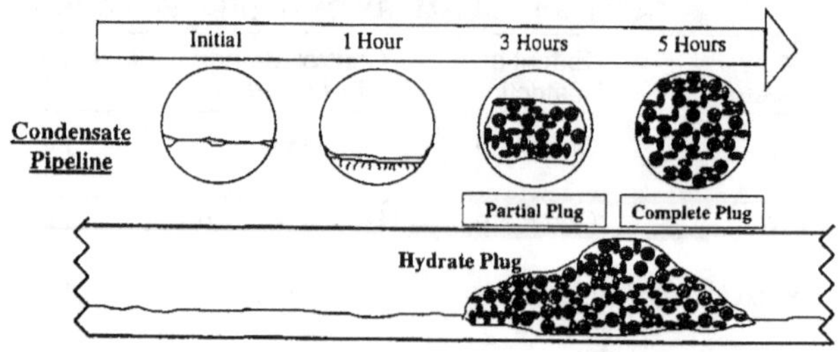

Fig. 3.2—Hydrate accumulation in condensate pipeline (Lingelem *et al.*[6]).

It often is very difficult to design and operate hydrate-free systems. If a line requires hydrate inhibition, it is very likely that hydrates will form over its lifetime. As such, it is prudent to identify likely points of hydrate formation so that hydrate prevention (or dissociation) can be addressed in the original design or in system operation through dehydration, heating, inhibitor injection, depressurization, or mechanical removal.

3.3 Techniques To Detect and Locate Hydrates

When partial or complete blockages are observed in flowlines, questions always arise about the plug location and composition. Where is the plug? Is the blockage composed of hydrates, paraffin, scale, sand, or some combination of these? Such questions are answered more easily with line access above ground (or above water) where a number of detection devices (e.g., thermocamera, gamma ray densitometers, or acoustic sensors) can be used as discussed in Sec. 3.3.1.

Indications of the blockage composition are obtained directly through combinations of separator contents and pig (sphere or ball) returns and indirectly as line-pressure drop. Separator contents and pig returns provide the best indication of pipeline contents and should be inspected regularly, even when blockages are not a problem. The separator discharge(s) and the pig trap provide valuable information about line solids, such as hydrates, wax, scale, and sand, and may be used as an early warning of future problems.

A less direct flow indicator is line-pressure-drop buildup, which differs for hydrates and for paraffins. Pressure-drop increases are usually more noticeable than flow-rate changes. With the exception of hydrate formation from gases as shown in Fig. 3.3a, hydrates usually cause a series of sharp spikes (Fig. 3.3b) in pressure as hydrate masses form, slough, agglomerate, and break before final blockage. With paraffins, the pressure buildup is gradual because deposition on the pipe wall causes a gradual increase in line-pressure drop (as in Fig. 3.3a). Clues to the type of plug can be found in pressure changes immediately before the blockage in addition to studying fluid slugging, gas/oil ratio (GOR), water cut, reservoir pressure, and choke setting, all of which can affect the flow and pressure drop.

When blockages occur in wells, it may be difficult to distinguish the cause. Frequently, only heating or mechanical means are available to detect the plug origin. In flowlines and wells, solid blockages of scale, rust, and sand are detected and removed less readily than hydrates or paraffins, so treatment for the more-solid plugs should be considered when hydrate and wax treatments fail.

In this section on detection of hydrate blockages, Sec. 3.3.1 considers early warning signs of hydrates and Sec. 3.3.2 considers methods to determine the center and length of the plug. This section relies primarily on Ref. 7 and information from the Statoil Hydrate Research/Remediation group, which contributed through in-depth interviews;* this group currently has more field experience in hydrate remediation than any other, perhaps by an order of magnitude.

3.3.1. Early-Warning Signs for Hydrates. Unfortunately, no one indicator gives the best warning of hydrate formation. Frequently, an increase in flowline-pressure drop, commonly thought to provide the best warning, is wholly inadequate for reasons given in the Early Warnings in Pipelines section. Instead, a suite of indicators should be used to provide the best early warning before blockages occur.

Of the three portions of the offshore process where hydrates form blockages, early indicators of well formation are the least developed. Hydrates in a well are announced most often by abrupt flow blockages accompanied by a high pressure drop. In normal operation, however, the well temperature is high enough to prevent hydrate formation. It is only during abnormal operations, such as start-up, shut-in, testing, and beginning gas lift, that hydrate formation

*Personal communication with T. Austvik, Statoil Research Center, Trondheim, Norway (13–15 July 1997).

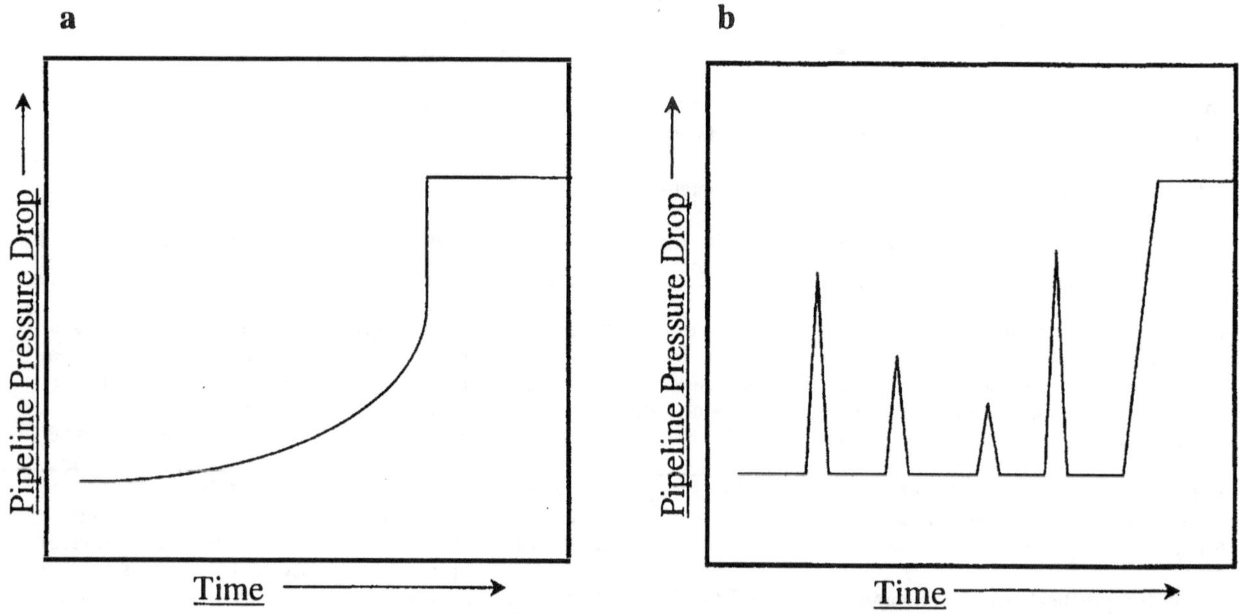

Fig. 3.3—Pipeline pressure drops because of hydrates: (a) hydrate buildup on walls and (b) hydrate-particle agglomeration.

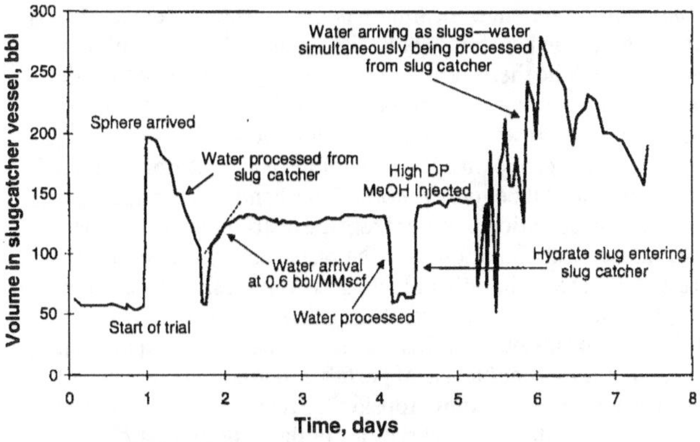

Fig. 3.4—Water production for wet gas line; DP=differential pressure (after Corrigan *et al.*[10]).

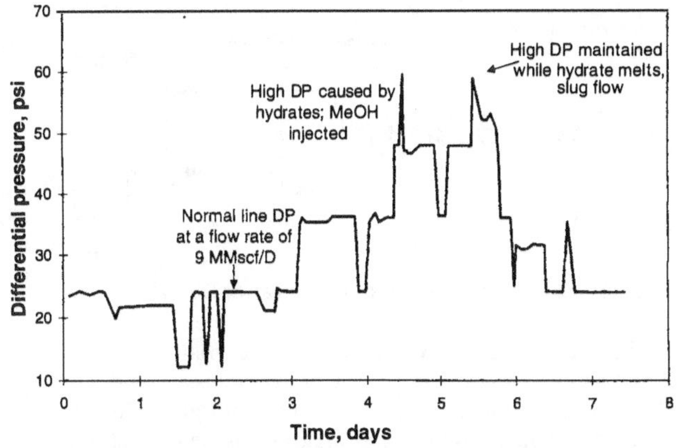

Fig. 3.5—Differential pressure caused by hydrate blockage (after Corrigan *et al.*[10]).

becomes a problem. Sec. 3.4, Techniques to Remove a Hydrate Blockage, provides information concerning when hydrates form without warning in a well.

Early-warning methods in the pipeline and onshore/platforms are discussed next. However, even with the methods listed in this section, there is a significant need for better hydrate detection.

Early Warnings in Pipelines. There are four methods for warnings of hydrate formation in a pipeline: (1) pigging returns, (2) changes in fluid rates and compositions at the separator, (3) pressure-drop increases, and (4) acoustic detection.

Pigging Returns. Periodically, flexible plastic balls or cylinders called pigs are pressure-driven through pipelines to clear them of condensed matter. A pig trip is initiated by means of a pig launcher and ended by a pig catcher or receiver, with the debris swept from the pipeline into a pig trap. Ref. 8 provides details of this technology.

Frequently, hydrate particles are found in pig traps before hydrate blockages occur in pipelines, signifying the need for corrective action (e.g., increased methanol injection). For instance, hydrate particles may occur when they have been suspended in an oil or condensate with a natural surfactant, such as the oil shown in Fig. 2.35 and discussed in Sec. 2.9. The Gullfaks subsea installation underwent several startups with hydrate present, but without problems before a blockage in January 1996.*

Rule-of-Thumb 18. A lack of hydrate blockages does not indicate a lack of hydrates. Frequently, hydrates form but flow with an oil/condensate (e.g., in an oil with a natural dispersant present) so they can be detected in pigging returns.

Pigging returns should be examined carefully for evidence of hydrate particles. Hydrate masses are stable even at atmospheric pressure in a pig receiver (or catcher) discharge. The endothermic process of hydrate dissociation causes released water to form ice, which inhibits rapid dissociation.[9]

Onshore pigging is commonly done to keep lines clear. However, it may be very expensive to provide offshore pigging either through a mobile pigging vessel over the well or from the wellhead without round-trip pigging capability. Such costs make examinations of offshore pigging returns an infrequent luxury.

Changes in Fluid Rates or Composition at Separators. When the water-production rate is small, it may be possible to monitor the rate of water production as an indication of hydrate formation. If the water arrival decreases appreciably at the separator, hydrates may be forming in the line.

Case Study 3.1. Separator Water Rate as an Indicator of Hydrate Production. In a controlled experiment, British Petroleum (BP) formed hydrates in a 14.5-in.-inner-diameter (ID), 13.7-mile-long gas line in the southern North Sea. Corrigan *et al.*[10] reported that, before the trial, water arrived at the separator in the amount of 1.3 bbl/MMscf. The test was started at the time marked Day 1 in **Fig. 3.4.** After methanol injection was stopped, the separator water arrival stopped completely after about 30 hours (no increase in accumulated water volume) while gas flow rates remained steady and the pressure drop did not change.

The first significant increase in line-pressure drop (to 35 psi in **Fig. 3.5**) was observed 46 hours after the start of the test. A further rise in p to 49 psi was noted after 3 days. Large fluctuations in the gas flow rate were observed 74 hours after the start of the trial, concurrent with further increases in p. A large slug of a condensed phase, presumed hydrates, arrived at the slug catcher at the trial conclusion. BP estimated 50 metric tons of hydrate were formed before methanol injection was resumed.

The previous case study is evidence that separator-water rate provides an early indication of hydrate formation by 16 hours in a gas line that has almost no oil/condensate and little water production. When water production is substantially higher or intermittent, it may be difficult to monitor changes in separator-water arrival for an early warning.*,**,[11]

Gjertsen[12] suggested that changes in gas composition provide an early indication of hydrate formation. In a rich gas field in the Norwegian sector of the North Sea, chromatograms showed a removal of hydrogen sulfide (H_2S) from sour gases as hydrates form. Hydrates particularly denude H_2S from natural gases owing to the near-optimal fit of H_2S in the small hydrate cavities.[3] The same is not true about the other acid gas, carbon dioxide.

Pressure-Drop Increases. The pressure drop (Δp) increases and the flow rate decreases if the pipe diameter is decreased by hydrate formation at the wall in a gas line. Because Δp in pipes is proportional to the square of turbulent flow rates, the change in Δp is more sensitive than the change in flow. With hydrates, however, a large restriction may be necessary over a long length before a substantial pressure drop occurs. For example, if a hydrate decreased the effective pipe diameter from 12 to 10 in. over a 1,000-ft section, the Δp would increase only 0.05 psi with 10 MMscf/D of gas at 1,000 psia and 39°F. In addition, the Δp trace usually contains substantial noise, making it difficult to observe small changes or trends in Δp.

Austvik† suggested that, while a gradual pressure increase in hydrate formation occurs for gas systems, a gradual pressure increase is not typical for a gas/oil/condensate system. In gas/oil/condensate systems, Statoil's experience is that, without advance warning, the line-pressure drop shows sharp spikes just before blockages occur. Fig. 3.5 shows the BP field experiment with methanol stoppage in a North Sea gas pipeline with little condensate or free water.[10] In that figure, step changes and spikes in Δp are more prevalent than a gradual increase.

*Personal communication with O. Urdahl, Statoil Research Center, Trondheim, Norway (14 July 1997).

*Personal communication with T. Austvik, Statoil Research Center, Trondheim, Norway (13–15 July 1997).
**Personal communication with J.L. Todd, 18 June 1997.
†Personal communication with T. Austvik, Statoil Research Center, Trondheim, Norway (13–15 July 1997).

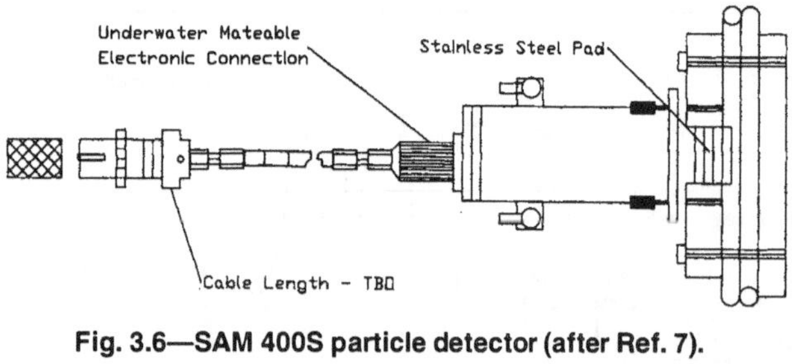

Fig. 3.6—SAM 400S particle detector (after Ref. 7).

In contrast, recent DeepStar Wyoming (onshore) trials show both gradual and spiked pressure drops in a gas-condensate field.[2] In Case Studies C.25 through C.27, the pressure built gradually upstream of a plug while pressure spikes indicated hydrate sloughing from the wall, with agglomeration and bridging downstream.

However, the onshore DeepStar tests had five pressure sensors spaced at intervals of a few thousand feet. As indicated in the previous calculation, with only two pressure sensors at either end of a line and because of the dampening effect of the gas, severe hydrate wall buildup must occur before a significant pressure drop is sensed. Most pipelines are likely to experience hydrates as sudden, extreme pressure drops.

Acoustic Sensing Along Subsea Pipeline. Ref. 7 states that "The only hydrate crystal detection instrumentation suitable for subsea use identified by this survey is sand monitoring instrumentation ...In a limited number of laboratory tests, an acoustic sand monitor has detected hydrates. However, a detailed study using the acoustic monitor has not been conducted."

Fig. 3.6 shows a typical acoustic sensor. More than 280 units have been installed to detect sand impingement on pipe by clamping the unit onto the flowline downstream of a 90° elbow or 45° bend. At flow rates as low as 3 ft/sec, the unit can detect 50 μm sand particles. Such units are rated for water depths of 4,000 ft and may be diver installed or installed by a remotely operated vehicle (ROV) with an underwater cable.

Acoustic sensors quantify the "hail-on-a-tin-roof" sound typical of hydrate particles impinging on a wall at a pipeline bend. Such sensors have ready application to hydrate detection in onshore pipelines. However, this unit has yet to be field tested in a subsea application. The initial background note of the Ref. 7 study presets a caution that still exists for hydrate detection, "This survey did not identify any proven hydrate or paraffin deposition measurement instrumentation for subsea multiphase flowlines or any other type of fluid-transmission lines. For gas-transmission lines, ultrasonic instrumentation has worked in specific applications and for single-phase liquid or gas lines, an acoustical/pigging system has been proposed."

Early Warnings in Onshore Processes and Platforms. In addition to the previous four types of subsea early-warning systems, two methods are suitable for detection of hydrates in an onshore system or on a platform, where piping and equipment are more accessible: thermocamera and gamma ray densitometer with temperature sensing.

Thermocamera. A thermocamera is a hand-held device that measures the infrared spectral transmission as an indicator of system temperature. The thermocamera is used only onshore or topside offshore with air between the detector and the suspected hydrate plug because water absorbs infrared transmission.

Statoil's hydrate group provided a hydrate-plug picture obtained by thermocamera, just beyond a short-radius bend in a topside riser (**Fig. 3.7**). Color-detection pictures from the instrument provide better temperature discrimination than the black-and-white reproduction presented here. While this blockage is obviously not an early warning, the picture is indicative of the instrument's ability.

As hydrate deposits build and as restrictions cause gas expansion, the low temperatures enable use of portable thermometers to detect plugs and potential plug points topside. A thermocamera enables determination of temperature variations in the system, particularly at points where hydrates might form but where a thermocouple typically is not provided, such as downstream of a valve.

The thermocamera is very sensitive to pipe coating, variations in wall thickness, and pipe roughness. After location of low temperatures, the engineer can determine the degree of intrusion of the system into the hydrate-formation region to consider corrective actions, such as insulation, heat tracing, and inhibitor injection.

Gamma Ray Densitometer With Temperature Sensing. A gamma ray densitometer uses an emitter and sensor on opposite, external pipe walls. The transmission of gamma rays to the sensor is a function of the density of the pipe contents. This technology is more than 50 years old and is used commonly in the chemical industry for level control in high-pressure, nonvisual systems.

Because densities of hydrates and water are very similar, gamma ray densitometry alone cannot discriminate between the two; at best, gamma ray measurements indicate changes in conditions that could be hydrates. In combination with the temperature downstream of the densitometer (such as at the platform startup heater in **Fig. 3.8**), hydrate formation can be discriminated.

Hydrates are indicated by a low temperature (perhaps owing to the Joule-Thomson effect) and an increase in density, whereas the water temperature is similar to that of gas. A high-density, low-temperature mass in the pipeline is likely to be hydrates, whereas a high-density slug without a temperature drop is probably water. As the blockage-removal section (Sec. 3.4) shows, even small pressure reductions cause hydrate dissociation, which results in heat being removed from the condensed phases, and lower temperatures. The temperature-sensing requirement makes it difficult to use the densitometer subsea, but it is a useful tool onshore.

3.3.2 Detection of Hydrate-Blockage Locations. The two objectives of locating the plug are to determine the distance from a reference point, such as an onshore separator or an offshore platform, and to determine the plug length.

In this section, three DeepStar reports were supplemented by Tommeliten field experiments.[1,2,7]

For onshore blockages, a direct method, such as the thermocamera (see Sec. 3.3.1), enables determination of location and length.

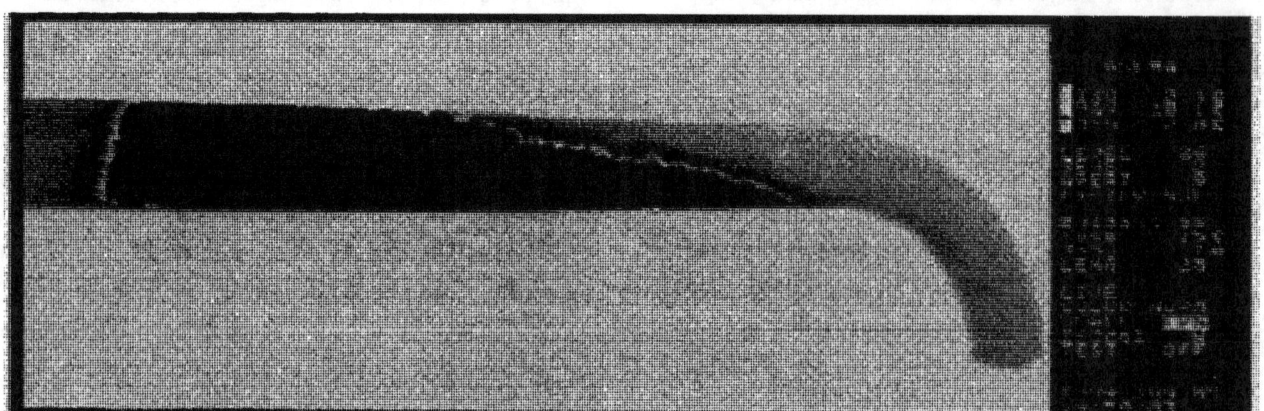

Fig. 3.7—Thermocamera picture of hydrates in horizontal portion of riser topside [personal communication with T. Austvik, Statoil Research Center, Trondheim, Norway (13–15 July 1997)].

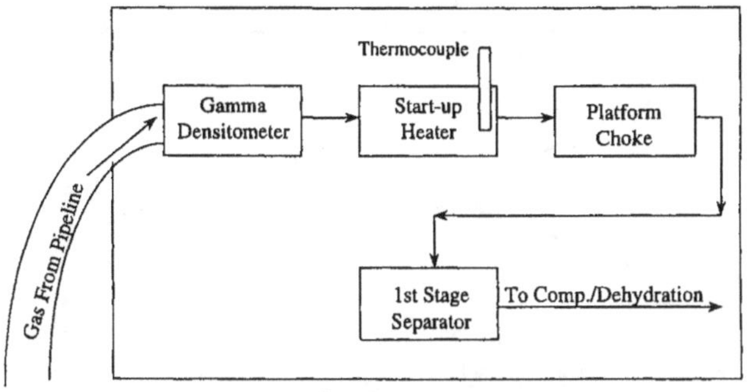

Fig. 3.8—Platform use of gamma densitometer.

For subsea offshore plugs, unfortunately there is no precise way to locate the blockage, so the methods involve both art and science. The efficiency of hydrate-blockage-location schemes is governed by the topology of the system, by the hydrate porosity shown in Fig. 3.2, and by Rule-of-Thumb 17 and its accompanying discussion.

The early-warning methods of Sec. 3.3.1 should be considered first. Additional methods to determine offshore hydrate-blockage locations are (1) filling the line/well with an inhibitor or using a mechanical/optical device; (2) reducing, or cycling the line pressure; and (3) measuring internal pressure through external sensors.

The Recommended Procedure To Locate a Hydrate Plug section, just before Sec. 3.4, gives a recommended composite blockage-location[5] method.

Inserting an Inhibitor or Mechanical/Optical Device in the Line/Well. When hydrates block an offshore flowline, it is common to fill the line with an inhibitor, particularly when the blockage is close to the platform. The blockage and the line topology may prevent the inhibitor flow from reaching a blockage far from a platform.

There is some disagreement about whether methanol or glycol should be introduced into the line as lubricant, and both are used. Because the methanol density is low, the higher-density glycol (and sometimes brine) is preferred.

The inhibitor injection volume allows the determination of blockage location relative to the platform, given the line size and a knowledge of liquid retention within the pipeline. In the following case study, the operator was fortunate to reach the hydrate plug with an inhibitor. In most cases, this method is ineffective.

Case Study 3.2. Methanol Lubrication Into an Export Line. Texaco reported a restriction in a 12.75-in. gas-export line from a platform at Garden Banks Block 189 in 725 ft of water. The export gas was dehydrated insufficiently and water condensed at a low point in the line, where hydrates rapidly formed.

The hydrate blockage was removed by venting from the platform before injecting methanol down the riser. Hydrates completely melted after a total of twenty to thirty 55-gal drums of methanol were used.

Rule-of-Thumb 19. Attempts to blow the plug out of the line by increasing the pressure differential results in more hydrate formation because higher pressures place the system farther into the hydrate-formation region. When a hydrate blockage is experienced in an offshore system, for safety reasons, the first step is to inject inhibitor into the line from the platform in an attempt to determine the plug distance from the platform. In an onshore system, inhibitor is injected from any access point.

Such a volumetric determination assumes that the plug is impermeable to the inhibitor and that the liquid holdup in the line is known (or negligible). Both assumptions may be incorrect because, before plugging, hydrate accumulations can push substantial liquids ahead of the plug.

Fig. 3.9—Wireline broach to determine hydrate location in well.

Case Study 3.3. Monoethylene Glycol (MEG) Lubrication Into Well Tubing. An operator experienced a blockage in a multiphase flow stream in the U.S. Gulf of Mexico, extending inside a deepwater riser connection between the platform and the seafloor, from 200 ft below to several hundred feet above the seafloor.

The well was being cleaned in preparation for production. The well contained 4- to 5-wt% calcium chloride completion brine. After hydrocarbon flowed from the well for a few hours, the well had to be shut in for 2 days because of bad weather, but methanol was not injected before shut-in. A gas-hydrate plug formed that held a differential pressure of 1,000 psi without movement.

Coiled tubing (see Sec. 3.4.4) was run down the tubing string, and ethylene glycol was jetted to remove the blockage. Jetting operations took 2 days, and the entire remedial operation required 1 week to complete.

For hydrates in a well, Statoil has used a broach, similar to that shown in **Fig. 3.9**, lowered on wireline to determine the blockage depth. Statoil also has used a similar wireline heating tool for hydrate dissociation in wells; in this case, the hydrate blockage can be located and dissociated with the same tool. Heating a hydrate blockage is not recommended unless the end is located because of safety reasons shown in Fig. 1.1 and its accompanying discussion. However, when the hydrate end is discernable, heating from one side of the blockage may be a viable option.

In a flowline, a wireline, a reach rod, coiled tubing, or fiber optics may be used to locate a plug. However, these detection methods are

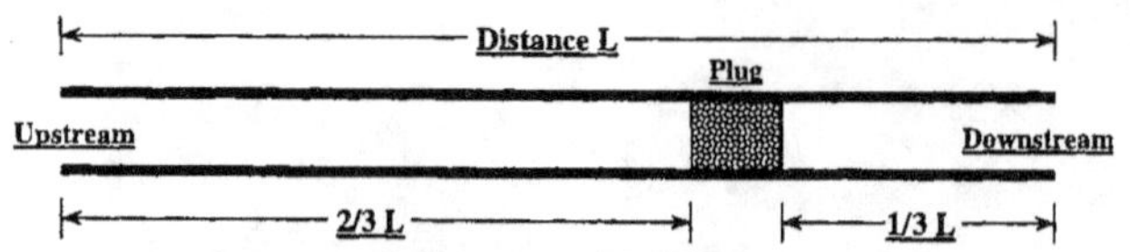

Fig. 3.10—Hydrate location in a pipeline.

currently limited to the first 10,000 ft from the platform and require mechanical intervention in the flowline.

Pressure Location Techniques. There are three pressure techniques to locate a hydrate blockage that are performed on the platform side of the plug: (1) pressure reduction, (2) backpressurization, and (3) pressure fluctuations. Each technique has advantages and disadvantages.

Pressure Reduction. This simple technique takes advantage of hydrate porosity and permeability to gas by decreasing the downstream pressure and monitoring the rate of downstream pressure recovery as well as the rate of pressure decrease of the upstream pressure of the plug. **Fig. 3.10** shows a flowline obstruction one-third of the way between the platform and the well. If the pressure is suddenly decreased and shut in downstream, the rate of downstream pressure recovery should be one-half the rate of upstream pressure decrease. With low-permeability plugs, patience may be required as the following case study illustrates.

Case Study 3.4. Depressurizing the Blockage for Location. In January 1996, a hydrate blockage occurred in a black-oil system in a 6-in.-ID, 1-mile-long line in the Gullfaks field. The normal oil rate was 18 Mscf/D, the water rate was 2,892 B/D, and the GOR was 100 to 360 scf/scf.

The normal line operating pressure was 2,420 psia, and the hydrate-equilibrium pressure (at the low temperature) was 261 psia. With the well shut in, the downstream pressure at the platform was rapidly reduced to 1,670 psia and then shut in. **Fig. 3.11** shows blockage upstream and the downstream pressure response (note expanded scale). Over a 25-hour period, the upstream pressure decreased approximately 73 psi while the downstream pressure increased the same amount. It was concluded that the plug was located midway in the pipe. Case Study 3.5 (Sec. 3.4) describes removal of this plug.

Two points should be emphasized about this case study. First, from a safety consideration, the pressure was reduced to a value just below the hydrate-dissociation pressure (not to atmospheric pressure) to prevent safety problems associated with a hydrate-plug projectile (see Chap. 1) propelled by a high differential pressure.

Second, pressure recovery was very slow, averaging only approximately 3 psig/hr. This slow rate might not be noticed if pressure is not monitored carefully by personnel, who may be inclined to discount slow changes relative to normal operation. The slow rate of pressure change was thought to result from the fact that most of the line contained liquid, which caused the apparent plug permeability to be about 1,000 times smaller than that for gas flow.

Statoil, the company with the most methodical, documented experience in hydrate remediation, prefers the previous method of plug location. The method locates the blockage center and the relative volumes upstream and downstream of the blockage(s). The disadvantage is that the method does not allow determination of the length of the blockage, the proximity of the blockage to the platform (because of the unknown plug permeability), or how multiple plugs may affect this location determination. The company locates the plug/platform proximity by inhibitor back-injection (see Rule-of-Thumb 19 and Case Study 3.2) or by backpressurization, as in the following method.

Backpressurization. One method to locate a complete pipeline blockage is to measure the pressure increase as metered amounts of gas are injected at the platform. The pressure-increase rate is proportional to the gas-input rate, enabling determination of the piping length between the platform injection point and the blockage.

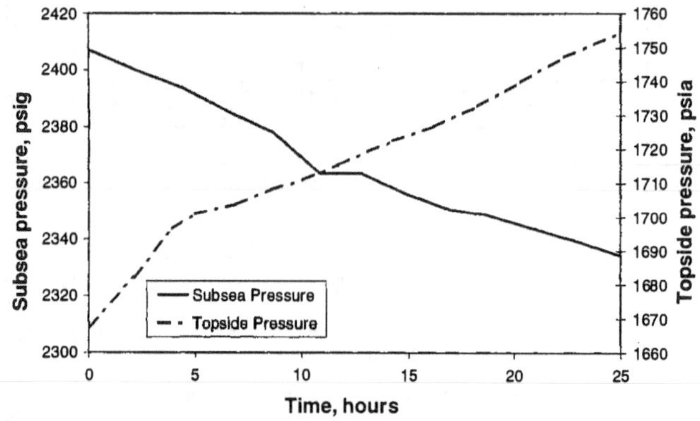

Fig. 3.11—Pressure change used to estimate plug location (after Gjertsen[12]).

Example 3.1: Backpressurization To Determine Plug Location. A 16-in.-ID gas pipeline is in full production when a hydrate plug occurs, blocking flow for a 0.6-gravity gas. The line is shut in, and the pipeline cools to the ambient temperature of 39.2°F. Before hydrate dissociation can begin to take place, the approximate location of the plug end should be obtained to determine the best remediation method and to evaluate safety concerns.

One standard location procedure is backpressurization. This method consists of pumping a known amount of gas into the pipeline from downstream and measuring the change in pressure over time. From the pressure increase, an estimate of volume can be obtained.

The following assumptions are made for the problem.

1. No porosity or permeability of the plug.
2. No liquid in the pipeline.
3. None of the injected gas condenses.
4. Constant temperature throughout the pipeline.
5. The heat of gas compression is dissipated rapidly.
6. The pipeline initially is at atmospheric pressure.

A reciprocating pump on the platform is used to inject gas at a rate of 4.89 lbm mol/min (1,850 scf/min) into the pipeline so that the pipeline pressure slowly increases. The heat of compression is assumed to be dissipated in the ocean and the entire temperature remains at 39.2°F.

The time required for the pipeline to attain even increments of pressure (e.g., 400, 600, 850 psia) is measured, and these data are used to estimate the pipeline volume downstream of the plug by

$$pV = ZnRT = V = \frac{ZnRT}{p}, \qquad (3.1)$$

where Z = gas compressibility as a function of p, T, and gas composition (obtained through an equation of state or from gas-gravity-compressibility charts[13]); n = value obtained from gas input as data (**Table 3.2**); p = corresponding pressure for n (Table 3.2); R = 10.73 (universal gas constant in units of psia, °R, lbm mol, ft^3); and T = seafloor temperature = 498.87°R.

TABLE 3.2—ESTIMATED PIPELINE VOLUMES BETWEEN PLATFORM AND PLUG

Data-Point Number	Time (minutes)	Pressure (psia)	Estimated Volume (ft^3)
1	60.76	400	3,637
2	96.58	600	3,664
3	144.39	850	3,667
4	198.61	1,100	3,662
5	300.01	1,500	3,663
	Average volume (plug to pipe end)		3,658

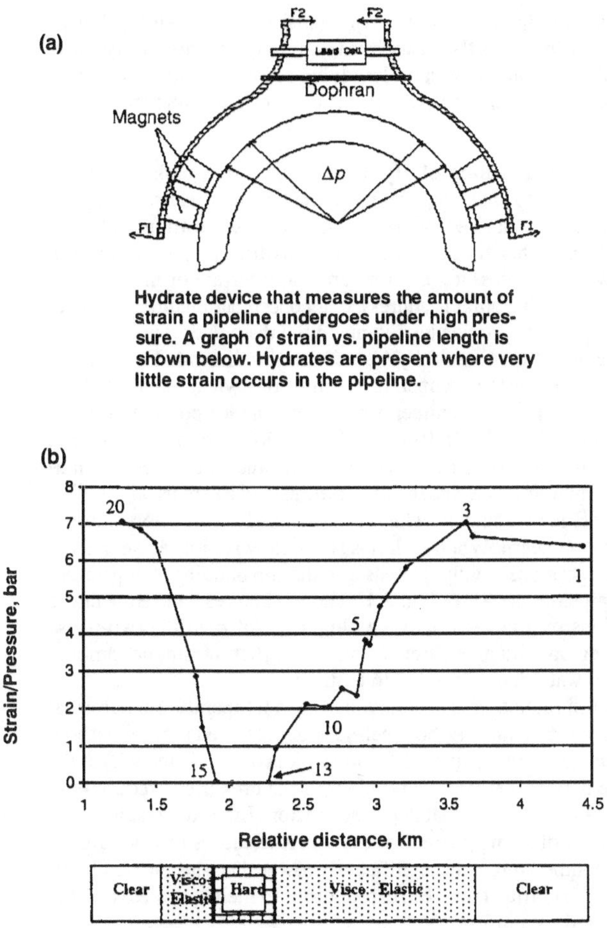

Fig. 3.12—Hydrate-plug detection through strain measurement; hydrates are present where very little strain occurs in the pipeline. (a) Hydrate device that measures amount of strain pipeline undergoes at high pressure. (b) Plot of strain vs. pipeline length. (After Ref. 1.)

Five data points provide a pipeline-volume average estimate between the hydrate plug and the end of the pipeline. A sample calculation for the first data point is as follows.

A line pressure of 400 psia is attained after 60.76 minutes when 297 lbm mol of gas has been pumped into the line. The gas compressibility is estimated at 0.915 from Figs. 23-7, 23-8, or 23-9 of Ref. 13. The pipeline volume is estimated as

$$V = \frac{ZnRT}{p} = \frac{(0.915)(297)(10.73)(498.87)}{(400)} = 3,637 \text{ ft}^3. \quad \text{(3.2)}$$

The first estimate of the pipe volume downstream of the plug is 3,637 ft^3.

Table 3.2 summarizes this same calculation for four other data points. The average for the pipeline volume downstream of the hydrate plug was 3,658 ft^3. The cross-sectional area of the pipeline is calculated to estimate the pipeline length between the plug and the pipeline end. The pipeline cross-sectional area is

$$A = \frac{\pi D^2}{4} = \frac{\pi 16^2}{4} = 201.06 \text{ in.}^2\left(\frac{1 \text{ ft}^2}{144 \text{ in.}^2}\right) = 1.396 \text{ ft}^2. \quad \text{(3.3)}$$

Because the pipeline volume equals length times cross-sectional area, the estimated location of the plug is 2,620 ft (= 3,658 ft^3/1.396 ft^2) away from the pipeline end.

Backpressurization has been implemented many times and is probably the method of choice for the majority of operators. However, there are several disadvantages that cause significant inaccuracies.

1. Because hydrate plugs frequently are porous (> 50%) and permeable, they transmit flow and act as a leak in a system calculated to be closed (i.e., no permeability).
2. The gas compressibility must be well-known to determine the pressure and volume rate increases.
3. The liquid holdup in the line must be known. This is particularly a disadvantage when significant elevation changes result in unknown liquid-holdup profiles or when the hydrate plug has accumulated liquid in front of it.
4. The location of multiple plugs cannot be addressed by this method; only the plug located nearest the point of gas injection can be determined.

Because of these inaccuracies, the method of backpressurization should be supplemented by other methods (see Sec. 3.3.2).

Pressure Variation. Refs. 1 and 7 discuss pressure-pulse travel time and pressure-frequency-response methods to locate a hydrate blockage. Both methods involve measurement of sound-wave travel time or frequency changes from the platform to the blockage. However, these analyses have not been successful to date because of two factors.

1. Acoustic response is a strong function of the relative amounts of gas and liquid, which are usually unknown and may occupy portions of a pipeline.
2. Reflected pulses are dampened by walls, valves, and bends and by a flexible plug.

Measuring Internal Pressure Through External Sensors. A technique was recently developed to measure hoop strain of the pipe as a function of line pressure to determine the location and type of blockage in an offshore line. An ROV places a metal caliper clamp on 25% of the pipe circumference with magnets (**Fig. 3.12a**). The platform end of the flowline is pressurized, inducing a hoop strain that is sensed by the pipeline caliper. The internal pressure causes a hoop strain that results in an outward movement of the caliper that varies with the wall deposits of the pipe. Lack of hoop strain across a section of pipe would indicate a blockage. The signal is transmitted to a work boat at the surface.

This method was used successfully in the North Sea on an 8-in., 9.32-mile-long flowline. **Fig. 3.12b** shows results of the strain gauge for 20 points at various lengths along a line blocked with paraffin. Points 13 through 15 are shown to be blocked with hard plugs, between viscoelastic plugs (Points 3 to 13 and 15 to 18) at either end. The map in Fig. 3.12b was in agreement with contents observed when the flowline was replaced. Deployment and recovery of each measurement required 1 to 2 hours. Because of difficulty of ROV deployment, this method has not yet been used to locate a hydrate pipeline plug.

Recommended Procedure To Locate a Hydrate Plug. There is no one precise method to locate a hydrate plug, so a combination of the previous methods are indicated next for best results. For example, in an above-ground line onshore, a thermocamera is recommended for locating a plug. For buried or subsea lines, a less rigorous procedure may be followed to locate the plug:

1. Estimate the hydrate-formation temperature and pressure of the blockage relative to the conditions of the pipeline. Use a simulation to determine where the pipeline contents enter the hydrate-formation pressure and temperature regime during normal operations. Confirm the simulation with a linear interpolation between the wellhead and platform temperature and pressure. This provides an approximation of the plug-initiation point; however, with flow blockage, the entire pipeline will cool into the hydrate-stability region. This calculation should be done during initial line design.
2. Depressurize the pipe downstream of the plug to approximately two-thirds of the pressure between the normal operating pressure and the hydrate-formation pressure. Do not decrease the pressure on one side of the plug below the hydrate-formation pressure. Monitor the pressure increase downstream and the pressure decrease upstream for the lesser time of either 24 hours or until a significant pressure change (e.g., 75 psig) is obtained at each point. Use the rate

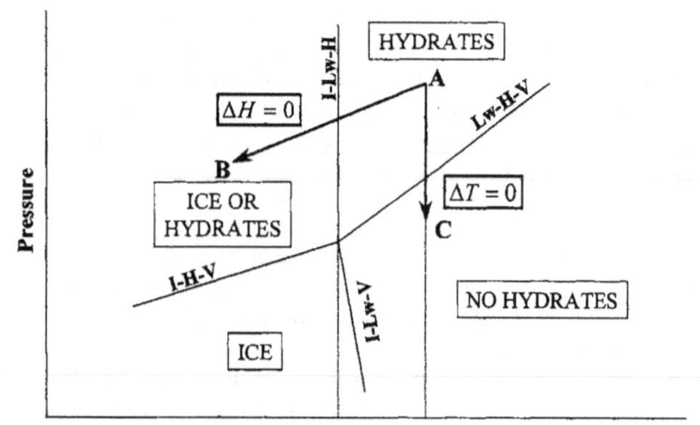

Fig. 3.13—Isenthalpic and isothermal plug dissociation (after Kelkar *et al.* [14]).

of pressure decrease upstream and increase downstream to determine the center point of plug(s), or relative volumes at each end of the plug(s).

3. In a subsea system, fill the riser with glycol to attempt to determine the distance between the platform and the plug. This may be inaccurate because of long distances to the plug, pipeline elevation changes, and partial line filling with condensate.

4. Backpressure the pipeline and monitor the pressure increase for a measured volume of gas input. Estimate the downstream distance from platform to plug by the pressure-change rate relative to gas input for a given compressibility and simulated liquid-retention volume. Use this technique with Step 2 to determine pipe volume before and after the plug.

5. If possible, use a mechanical device to determine plug location.

3.4 Techniques To Remove a Hydrate Blockage

Four techniques to remove a hydrate blockage are listed in order of frequency of success.

1. Hydraulic methods, such as depressurization (Sec. 3.4.1).
2. Chemical methods, such as injection of methanol or glycol (Sec. 3.4.2).
3. Thermal methods that involve direct heating (Sec. 3.4.3).
4. Mechanical methods with coiled tubing and drilling (Sec. 3.4.4).

Applications of these methods can be divided further into three cases: (1) partial blockage, (2) total blockage without a liquid head, and (3) total blockage with a liquid head. The following discussions concern only the final two cases. It is assumed that any indication of a partial blockage will be treated promptly with massive doses of methanol, the most effective inhibitor. Combinations of removal methods usually are tried simultaneously, making it difficult to assess the individual-method effectiveness.

Rule-of-Thumb 20. Regardless of the method(s) used, the time required for hydrate dissociation is usually days or weeks. After a deliberate dissociation action is taken, both confidence and patience are required to observe the result over a long period of time.

It is often suggested that corrective actions be changed almost hourly when immediate results are not observed. Rapidly changing corrective actions are futile without observations of significant effects on plug removal. The waiting aspect of plug removal is frequently the most difficult for operating and engineering personnel, who are accustomed to producing results on a continuous basis. Typical times of days or weeks are required for plug removal, as indicated by Appendix C case studies. Measurements, such as pressure drop across the plug, are monitored and changed continuously, but only after sufficient time to assess the current method.

Rule-of-Thumb 21. When dissociating a hydrate plug, it always should be assumed that multiple plugs exist, from both safety and technical standpoints. While one plug might cause the initial blockage, a shut-in causes the entire line to cool rapidly into the hydrate-formation region and several low-lying points of water accumulation convert rapidly to hydrate at the water/gas interfaces.

3.4.1 Depressurization of Hydrate Plugs. From both safety and technical standpoints, the preferred method to dissociate hydrate plugs is to depressurize from both sides. Depressurization is particularly difficult when the liquid head on the hydrate plug is greater than the dissociation pressure, as in mountainous terrain or in very deep water. Before that point is addressed, a conceptual picture of hydrate plugs is given to provide key points in the dissociation process.

Conceptual Picture of Hydrate Depressurization. **Fig. 3.13** illustrates the pressure/temperature conditions when a hydrate plug occurs in a pipeline at temperatures above the ice point. To the left of the three-phase (L_w-H-V or I-L_w-V) lines hydrates or ice can form, while to the right only fluids can exist. Because the lowest ground-burial temperatures or ocean temperatures (39°F) are usually higher than 32°F, ice formation (which also blocks flows) is not a normal operating concern. When hydrates form, flow is blocked so that the plug temperature rapidly decreases to the surroundings temperature at the pipeline pressure. Fig. 3.13 shows the pressure/temperature conditions of a pipeline hydrate plug at Point A in the two-phase (H-V) region with an excess gas phase, so that the smaller amount of liquid water has converted to hydrate.

Pressure reduction is accompanied by a temperature decrease at the hydrate interface. If the pipeline is depressured rapidly without heat transfer, Joule-Thomson (isenthalpic) cooling (Line AB) at the hydrate may exacerbate the problem. If the pressure is reduced extremely slowly, isothermal depressurization (Line AC) results. Usually, an intermediate pressure-reduction rate causes the hydrate-interface temperature to be significantly lower than the surroundings, causing heat influx from the surroundings to melt hydrates from the pipe boundary inward.

With rapid or extreme pressure reduction, the hydrate-equilibrium temperature can decrease to less than 32°F for a methane hydrate depressured to atmospheric pressure. In this case, the water from the dissociated hydrate buffers the temperature reduction to approximately 0°C, by converting to ice below the solid/liquid line (I-L_w-H shown in Fig. 3.13). If ice formation occurs with hydrate dissociation, the question arises of how the ice-plug dissociation rate compares with the hydrate dissociation rate in an ocean pipeline.

Recent experiments[14] and modeling suggest that blockages are removed most efficiently when the line is depressurized to the fullest extent as rapidly as possible. When ice forms, it has a lower temperature and higher thermal diffusivity than hydrates, resulting in a rate increase of heat transfer into the pipe.

In 1994–97 field studies, more than 20 hydrate plugs were systematically formed and removed from a 6-in. North Sea line in the Tommeliten Gamma field.[11,12] In both laboratory and field studies, these plugs were found to be very porous (> 50% porosity) and permeable. Porous, permeable hydrates transmit gas pressure easily while still acting to prevent free flow in the pipeline. When the pressure was decreased at both ends of a highly porous hydrate plug, the pressure decreased throughout the entire plug to a constant value. The dissociation temperature at the hydrate front is determined by the pipeline pressure and by the buffering capacity of the water fusion to ice.

Pipeline depressurization reduces the hydrate temperature below the temperature of the surroundings. Heat flows radially into the pipe, causing dissociation first at the pipe wall (**Fig. 3.14**). Radial hydrate dissociation controls plug removal because the pipe diameter (typically less than 2 ft) is at least an order of magnitude less than the length of a hydrate plug (frequently more than 50 ft) in a pipeline.

The radial-dissociation concept contrasts with previous longitudinal-dissociation concepts of nonporous hydrates where depressurization from both ends was thought to result in dissociation progressing from the plug ends toward the middle.[1,15] When the temperature of the hydrate is lower than that of the surroundings, heat flows radially into the plug, causing dissociation along the entire length (**Fig. 3.15**). Some plug dissociation occurs at the ends,

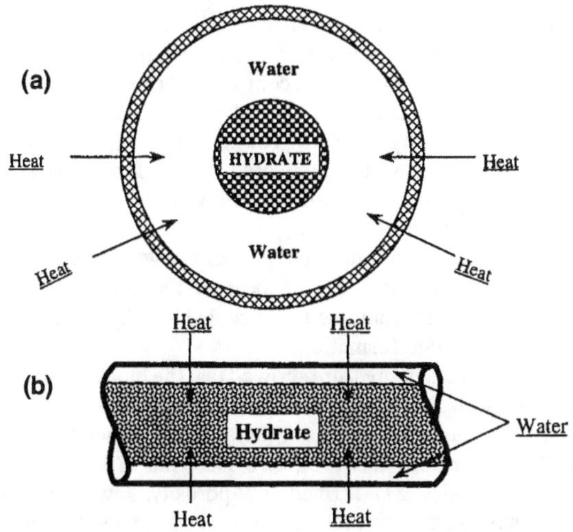

Fig. 3.14—Radial dissociation of hydrate plug: (a) pipeline cross section and (b) pipeline longitudinal view.

but, owing to much smaller dimensions, the radial dissociation (which occurs simultaneously along the plug length) controls blockage removal.

Fig. 3.15 shows a cross section of a pipeline hydrate plug that has been depressured to provide an equilibrium temperature just above 32°F. Such a pressure corresponds to approximately 450 psia for a pure methane gas, as low as 180 psia for natural gas, as predicted by the methods of Secs. 2.6. and 2.7. Fig. 3.15a shows an inner hydrate core enclosed in a water layer, which results from hydrate melting. The water layer is adjacent to the pipe wall. Fig. 3.15b shows the temperature profile from a deep ocean temperature of $T_o = 40$°F at the pipe wall to T_H, the hydrate-dissociation temperature (set by the line pressure to a point just above the ice point), where it remains uniform throughout the hydrate layer. As a result, the radial disappearance of the two-phase water/hydrate boundary (X_1) determines the disappearance of the final solid and eliminates the flow obstruction.

Because hydrate-plug detachment occurs first at the pipe wall, a partially dissociated plug moves down the pipeline when the line is restarted but may later result in a plug at a pipeline bend, depression, or other obstruction. The second blockage by the plug can be more compact than the first if, for example, there is substantial momentum on impact at the bend. In the extreme, the plug acts as a projectile that may result in the severe safety problems indicated in Chap. 1. This phenomenon relates to Rule-of-Thumb 20, indicating that one of the most important aspects of plug removal is patience to allow time for total dissociation.

The concepts are similar for both onshore and subsea pipelines. In the previous conceptual picture, it was assumed that the pipeline-wall temperature is constant at 40°F. If a line is insulated, hydrate dissociation becomes much more difficult because the insulation that prevented heat loss from the pipe in normal operation now prevents heat influx to the pipe for hydrate dissociation. Alternatively, if the pipe is buried, the pipe-wall temperature will be greater than 40°F and the system may have some insulation from the ground.

The cross section in **Fig. 3.16a** shows a hydrate-plug removal when the dissociation pressure is much lower. An inner hydrate core is surrounded by an ice layer that is enclosed in a water layer adjacent to the pipe wall. **Fig. 3.16b** shows the temperature profile from $T_o = 40$°F at the pipe wall, to $T_I = 32$°F at the water/ice interface, to a lower hydrate-dissociation temperature, T_H (set by the line pressure), at the ice/hydrate interface, where it remains uniform throughout the hydrate layer. As a result, there are two two-phase boundaries: a slowly dissociating water/ice boundary (X_1) and a second, rapidly dissociating ice/hydrate boundary (X_2). We are particularly interested in the rate of progress of X_1, which is slow enough to determine the disappearance of the final solid (ice), because any solid phase constitutes a flow obstruction in a pipeline.

Hydrate dissociation at a low pressure almost always results in ice formation. Hydrate removal is accomplished by both depressurization and heat influx from the surroundings, while ice-plug removal must rely on heat influx alone. However, because the ice is at a lower temperature and has a higher thermal diffusivity, the entire solid mass dissociates more rapidly by converting to ice (at low dissociation pressure) than by maintaining a high pressure so that the temperature is above the ice point, as in Fig. 3.15.

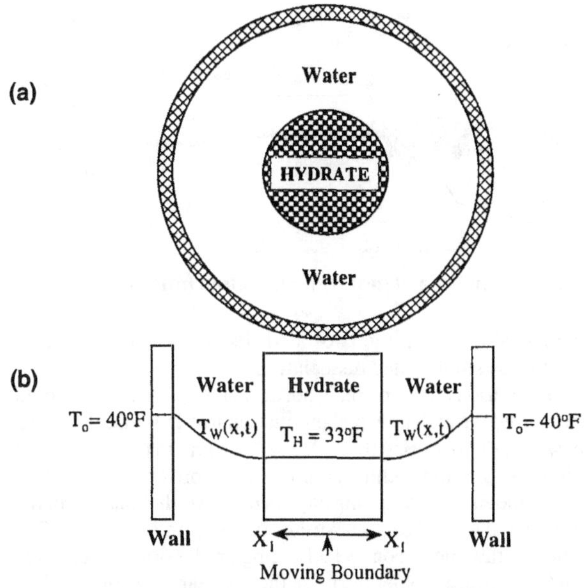

Fig. 3.15—Hydrate dissociation with water present: (a) pipeline cross section and (b) pipeline temperature.

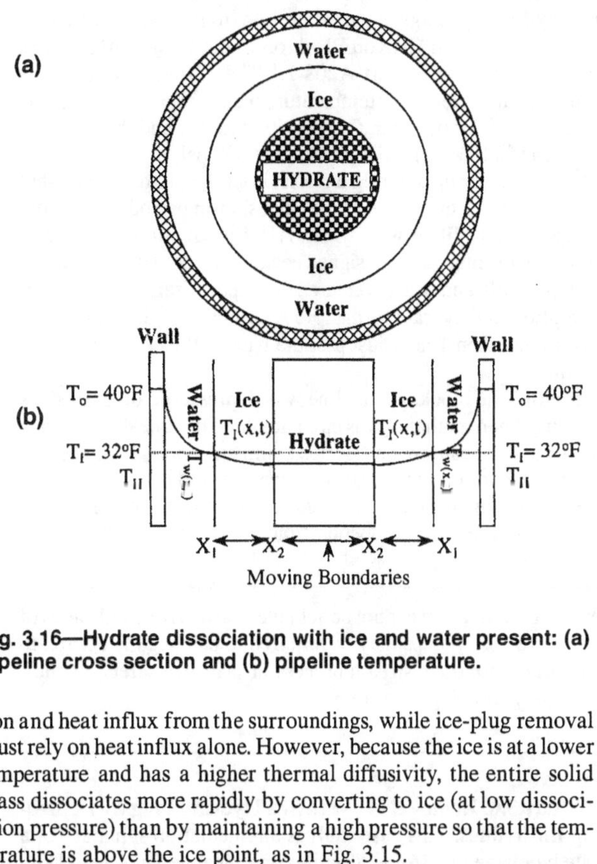

Fig. 3.16—Hydrate dissociation with ice and water present: (a) pipeline cross section and (b) pipeline temperature.

Austvik[11] noted some exceptions to radial dissociation, particularly for plugs of low porosity/permeability or for very long plugs. Berge *et al.*[4] show that hydrate plugs consolidate with time after plug formation, causing porosity and permeability to decrease considerably. Hydrate plugs should be dissociated as soon as possible to take advantage of higher porosity and permeability.

Hydrate Depressurization From Both Sides of Plug. There are two reasons for the preferred method of two-sided hydrate-plug dissociation.

1. For a single plug, dissociation from both sides eliminates the safety concern of having a projectile in the pipeline.

2. Two-sided dissociation eliminates Joule-Thomson cooling, which may stabilize the downstream side of the plug. With radial

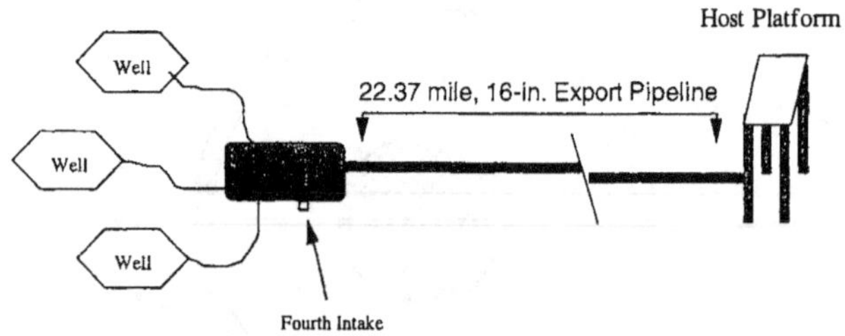

Fig. 3.17—Offshore platform and manifold (after Lynch[16]).

dissociation along the plug, two-sided dissociation is more than twice as fast as single-sided dissociation.

For these reasons, the upstream portion of a hydrate plug should be depressured through a second production line if available. If this is impossible, depressurize through a service line manually used to inject inhibitors at the wellhead; in this case, provision should be made for removing or bypassing any check valve that may be in the service line at the wellhead. In extreme cases, as in Case Study 3.6, it may be worthwhile to connect a floating production vessel to the manifold or wellhead to depressurize the upstream side of the plug.

Case Study 3.5. Gulf of Mexico Plug Removal in Gas-Export Line. A hydrate blockage in the export line from Shell's Bullwinkle platform in the Green Canyon Block 65 to the Boxe platform was reported in DeepStar Report A208-1.[1] The 12-in., 39,000-ft line was not insulated. Seawater temperature was 50°F at the base of the platform in 1,400 ft of water. Gas gravity was 0.7, and the flow rate was 140 MMscf/D at an inlet pressure of 800 psi.

Gas hydrates formed during a restart after the platform was shut down because of a hurricane. During the shut-in period, the gas dehydrator partially filled with water. After production restarted, it was not dehydrating gas as designed because the dehydrator was not cleaned properly and wet gas entered the export riser, causing water condensation and hydrate formation. A complete hydrate blockage formed in less than 1 hour just past the base of the export riser at a low spot.

To remove the blockage, the line was depressured on both sides of the plug. Then methanol was circulated into the line to accelerate the hydrate-dissociation rate. After complete removal of the hydrates, the dehydrator was cleaned, inspected, and restarted properly. The entire remedial operation required 36 hours to complete. The major cost was the lost production time.

When depressurization cannot be achieved easily from both sides of a plug, more costly steps may be required to balance the depressurization to ensure downstream process or platform safety, as indicated in the following case study.

Case Study 3.6. Removal of North Sea Hydrate Plug by Depressuring Both Sides. This case study is a remediation summary of a hydrate blockage in a 16-in., 22-mile-long pipeline between a U.K. sector North Sea gasfield well and platform.

Plug Formation. *Setting.* The gas field was located in the southern North Sea and consisted of three subsea wells, flowing into a subsea manifold with a capacity of four well inputs. **Fig. 3.17** is a schematic of the manifold line and platform. The gas composition, temperature, and pressure promoted hydrate formation, consequently MEG was injected into the manifold and wellheads to inhibit hydrates thermodynamically. The inhibited water, gas, and condensate was pumped through a 22-mile, trenched, insulated export pipeline to a processing platform where water was removed from the condensate. The MEG in the pipeline was recycled and piped back to the manifold by means of a 3-in. pipeline piggybacked to the export line.

Blockage. On 14 April 1996, an unusually large liquid slug overran the platform primary separator, causing a temporary shutdown. The liquid slug was removed, but complete blockage of the pipeline had occurred during shutdown. It was hypothesized that the blockage was a result of a hydrate plug for the following reasons.

- The pipeline free water, recovered during depressurization at the platform, did not contain MEG inhibitor. The 3-in. MEG inhibitor piggyback line had ruptured.
- Through backpressurization, the blockage was found to be 492 ft away from the platform. At this location, the pipeline exited the mudline, allowing its contents to be cooled rapidly by ocean currents, causing hydrate formation.
- Slight decreases in pressure showed that the blockage had some porosity. This also had been observed for several Statoil hydrate plugs (see Tommeliten field Case Studies C.15 through C.17 in Appendix C. In contrast, however, two DeepStar field trials (Case Studies C.26 and C.27) formed low-porosity, low-permeability plugs that would transmit pressure very slowly and withstand high pressure drops.
- The liquid slug that shut down the compressors probably was caused by a partial hydrate plug propelling a fluid front down the pipeline.

The hydrate-blockage proximity to the platform posed serious safety concerns. Pipeline depressurization on both sides of the hydrate plug was necessary to dissociate the hydrate. If only the downstream side of the blockage was depressured, the pressure differential might have caused a projectile to form that could destroy the riser piping and damage the platform. The projectile could be life threatening to workers on the platform and result in costly damages to the platform itself.

Consequently, depressurization had to be done through both the platform and the subsea manifold to ensure safety. Projectiles could form because of dissociation if gas became trapped within multiple plugs. Slow depressurization was required to remove pressure buildups in the hydrate plug(s). Several methods were considered.

Depressurization Method. *Initial Ideas.* Three questions were raised to determine a proper depressurization method.

1. Will the remediation process effectively remove the obstruction?
2. What is the cost of equipment and modifications?
3. How much time is needed to complete the remediation?

On the basis of these questions, process engineers, consultants, safety management, and diving specialists proposed three potential depressurization methods.

1. Jackup Rig.

Method: Tow a jackup rig to the site. From the rig, attach a high-pressure riser to the manifold's subsea tree and flare exiting gas through the rig's flare stack.

Modification: A spool piece would have to replace a nonreturn valve on the manifold's fourth well intake.

Time Required: A drilling rig was not currently available, consequently a delay of approximately 8 weeks was needed to locate a suitable rig. The time required for hydrate removal could be 12 weeks.

Estimated Cost: U.S. $1,980,000.

Feasibility: The large amount of time required to locate a jackup rig made this an ineffective remediation method, useful only in the absence of other methods.

2. MEG Injection Line.

Method: Connect the subsea manifold's spare fourth flange to the 3-in. MEG injection line and flare gas at the platform.

Modification: Subsea work would require a spool piece installed between the two pipelines. An additional line to inject inhibitor would be needed to prevent MEG-injection-line hydrate blockage. The platform (while in operation) required significant modification to connect the MEG pipeline to its flare stack. To complicate the matter further, all the MEG currently in the pipeline would require storage on the platform, which had limited space.

Time Required: 6 to 8 weeks.

Estimated Cost: Unknown, expected to be higher than the other methods because of the numerous modifications that were required.

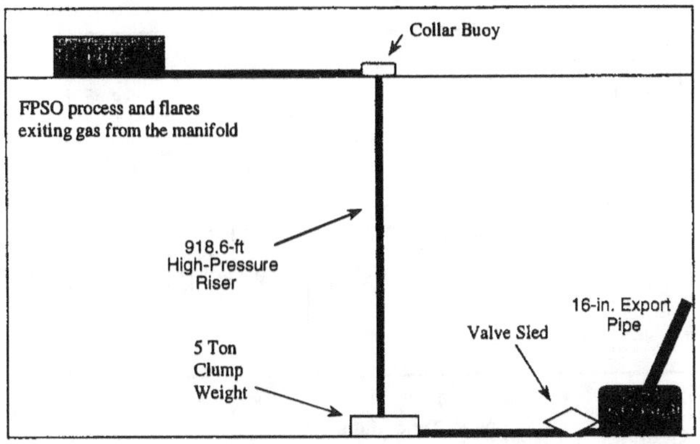

Fig. 3.18—Preliminary remediation setup (after Lynch[16]).

Feasibility: Both substantial platform modifications and the question of flow assurance through the MEG injection line made this remediation method impractical.

3. Floating Production and Storage Offloading (FPSO) Vessel. Method: Connect an FPSO with a processing plant and flare to the subsea manifold's fourth nozzle to depressurize, process, and flare the gas upstream of the plug. The connection between the manifold and FPSO would be made through a high-pressure, flexible riser.

Modification: The platform required no modifications. A diving rig was required for the subsea work. A valve skid also was needed; this contained both emergency shutdown valves (ESDV's) and an MEG injection valve. The flexible riser and the manifold would be connected with a spool piece. **Fig. 3.18** is a schematic of the connection design.

Time Frame: An FPSO was available for immediate use, consequently the required time was expected to be 6 to 8 weeks.

Estimated Cost: U.S. $1,906,000.

Feasibility: This method proved to be the most feasible. The immediate availability of a FPSO and diving rig allowed modifications to begin. It was estimated that the FPSO could be at the site and begin within 2 weeks.

Establishing Procedures/Permits. It took approximately 2 weeks to develop remediation alternatives. Procedures were then written to establish the pipeline depressurization process firmly, taking into account safety and coordination requirements between the diving rig and the FPSO. All parties were educated about the tasks involved.

Government permits were applied for at the Health and Safety Executive Pipeline Inspectorate and the Dept. of Trade and Industry Oil and Gas Office for additional gas-flaring and well modification. The permits were expedited by local agencies to prevent delay in hydrate removal. Two weeks were required to prepare procedures and permits for depressurization. In the meantime, the FPSO and diving rig were being equipped for the operation and were moving to the field.

Depressurization of the Pipeline. *Operations.* The diver's first task was to locate the subsea manifold's fourth nozzle manually and to isolate it from any trees or flow loops. The fourth nozzle then was modified with a spool piece for connection with the high-pressure riser. The valve skid was now ready to be put in place. Because of the sandy ocean bottom, it became necessary to provide a foundation for the valve skid. The valve skid was placed on a concrete mattress and then stabilized with gravel bag supports coupled with Tirfors, chain blocks, and ground anchors. This ensured that no movement would transfer from the flexible riser to the valve skid. The valve skid contained ESDV's and an MEG injection system for the pipeline. **Fig. 3.19** shows subsea valves and manifold attachment.

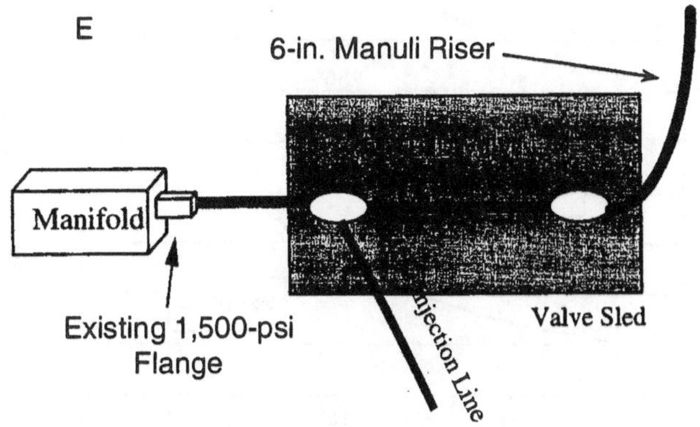

Fig. 3.19—Valve sled with manifold interface (after Lynch[16]).

The diving-rig crew then inspected the flexible-riser route to ensure that is was clear of debris. It proceeded to deploy 920 ft of the high-pressure riser by a tugger rigged with a dead-man's anchor. The MEG in the riser provided some buoyancy, consequently the line was anchored through concrete mattresses. A 5-ton clump weight was placed at the riser bottom, with a buoyancy collar attached to the surface.

The FPSO could process gas only at 600 psig, consequently it required some modification to process the 1,300-psig pipeline gas. Additionally, a quick-release valve (QRV) was needed to enable the FPSO to escape from the riser in an emergency; however, available QRV's could not withstand pressures of 1,300 psig. Initial design placed choke valves in the riser to reduce pressure for the QRV; however, this caused control problems and was deemed impractical.

A new, innovative QRV was developed with a standard-valve weak link with three additional hydraulic jacks for manual release. This valve could withstand 1,500 psig of pressure, allowing choke valves to be placed on the ship's deck, which simplified control issues. This design enabled a safe, simplified control of gas pressures from the deck of the FPSO. **Fig. 3.20** describes the system.

The buoyancy of the riser prohibited pipeline intake through the FPSO's moonpool. Spool pieces were used to allow riser intake from the side of the ship deck. The riser was also steam traced with 1,000 ft of 1-in. piping to maintain the minimum processing temperature required by the FPSO.

Fig. 3.21 is a complete picture of the FPSO attachment to the subsea manifold. All valves and risers were tested and shown to be in

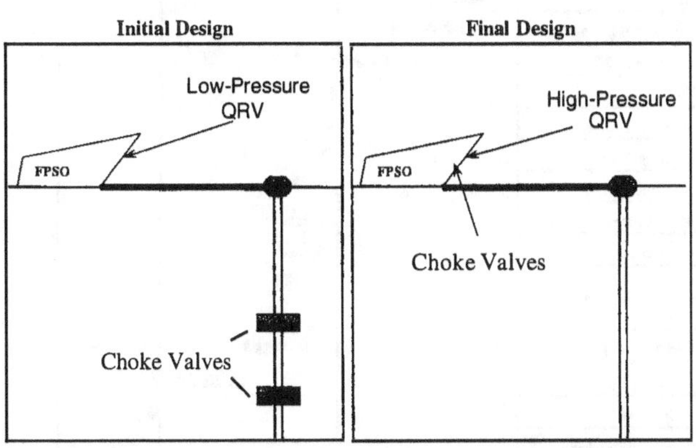

Fig. 3.20—Design with high-pressure QRV, which allowed placement of choke valves on FPSO deck and eased pipeline control (after Lynch[16]).

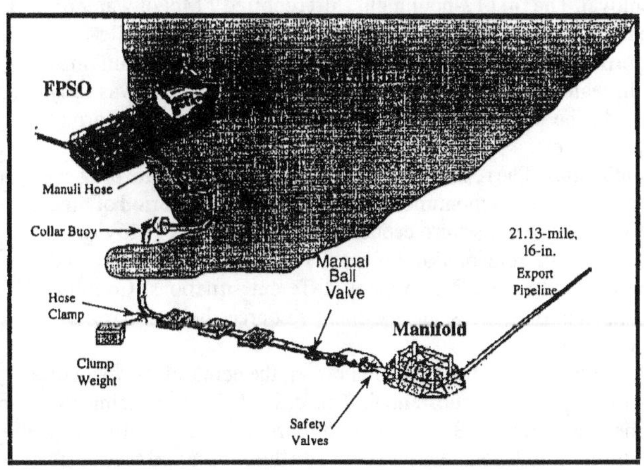

Fig. 3.21—Complete FPSO/manifold interface (after Lynch[16]).

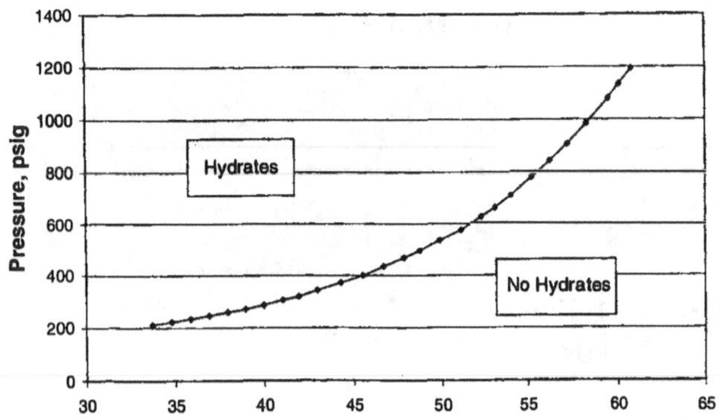

Fig. 3.22—Hydrate-formation curve (after Lynch[16]).

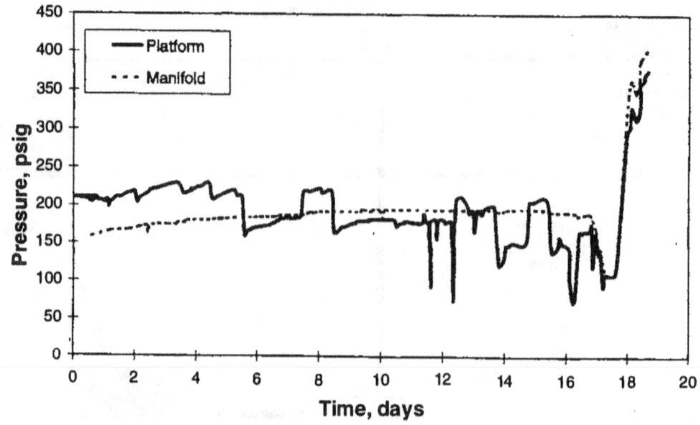

Fig. 3.23—Pressure of manifold and platform during hydrate remediation (after Lynch[16]).

working order. Overall the modification and installment procedures required 1 week before pipeline depressurization could begin.

Determining the Pipeline Minimum Pressure. Reducing pipeline pressure too much could have resulted in ice formation, which was thought to require significantly more time than hydrate dissociation. Ice formation was prevented through use of the hydrate equilibrium curve (**Fig. 3.22**) for the field.

At constant low pressure, hydrates dissociate continually, maintaining the equilibrium temperature at that given pressure. As the graph illustrates, the equilibrium pressure at 32°F was 200 psig. Consequently, the FPSO reduced the pipeline pressure to 185 psig to maximize hydrate dissociation while minimizing ice formation.

Depressurization. Twenty-three days were required to dissociate the pipeline hydrate completely. Heat transfer between the ocean and the pipeline was slow because the line was trenched and insulated in the seafloor. Dissociation was slightly facilitated by occasional backpressuring, which drew methanol into the plug. Backpressuring also proved beneficial in determining the location of the plug. **Fig. 3.23** shows the pressures in the pipeline throughout the depressurization process. Note the slight pressure increases that occurred during depressurization. These formed as a result of gas pockets that were released suddenly as portions of the plug(s) dissociated.

The pressure was monitored for 12 hours after the hydrate was thought to be dissociated. No pressure variation was noticed so the flexible riser was recovered and the depressurization apparatus dismantled. Throughout the whole operation, no equipment failure occurred and the operation progressed smoothly.

Recommissioning the Pipeline. After the hydrate was dissociated, significant amounts of free water remained in the pipeline, causing careful recommissioning to prevent reformation of hydrates. Above-normal amounts of MEG were added to the system before pipeline startup. One gas well was opened with flow to the platform to maintain low pressure, preventing hydrate formation. The high platform intake caused a high gas velocity that facilitated rapid water removal. The first 12-hour night shift reported 7 Mcf of water received from the separator, which resulted from a 1.25-mile-long (nonporous) hydrate plug. The high flow rate of gas was maintained until the water contained 40% MEG, ensuring that the line was fully inhibited. The pressures and intakes then were returned to normal operating levels.

Conclusions. The remediation team removed the hydrate plug efficiently, achieving a monumental task in a very short period of time and preventing more severe economic losses. **Fig. 3.24** provides a timetable of the remediation process. The procedure and method followed could be applied to many different situations. Communication, clear objectives, and excellent resources helped in removing the hydrate plug.

Despite the efficient remediation effort, the economic impact of the hydrate plug was substantial. The cost of depressurizing the pipeline was almost U.S. $3 million dollars, without counting lost production. In addition, relations between the buyers and producers were tested because of lack of production. Fortunately, good initial relations between the two reduced the impact of the disruption. This case study shows the potential financial loss that can result from hydrate plugs. Hydrate prevention is key in preventing significant economic and production losses.

Hydrate Depressurization From Both Sides of Plugs With Significant Liquid Heads. Results similar to those of Case Studies 3.5 and 3.6 may not be applicable with drastic changes in elevation, such as mountainous terrain or very deep ocean plugs. When depressuring a multiphase pipeline with large elevation changes, the hydrostatic pressure or liquid head against the plug face(s) may be higher than the hydrate-dissociation pressure. However, the removal of liquids from each side of a hydrate plug may be difficult.

To date, little documented experience exists for depressuring plugs with substantial liquid heads in deepwater lines. However, the situation has been evaluated in light of most of the case studies in Appendix C, and Example 3.2 provides recommendations.

Example 3.2: Methods of Liquid-Head Removal in Plugged Lines. This example abstracts an in-depth study of fluid removal as a preliminary step to depressuring lines.[1]

Fig. 3.25 shows the L_w-H-V equilibrium conditions for the Hercules and Jolliett fluid conditions in a 50-mile pipeline in 4,000 ft of water in the Gulf of Mexico. When a blockage occurs, if the gas is not vented, the temperature rapidly decreases to 40°F with a pressure between 2,000 to 3,000 psia (a subcooling of 30 to 33°F). After gas venting, the pressure is 1,000 to 1,300 psia, a factor of five to six times greater than the equilibrium pressure (200 psia) at the ocean-floor temperature (40°F) with a subcooling of 22°F.

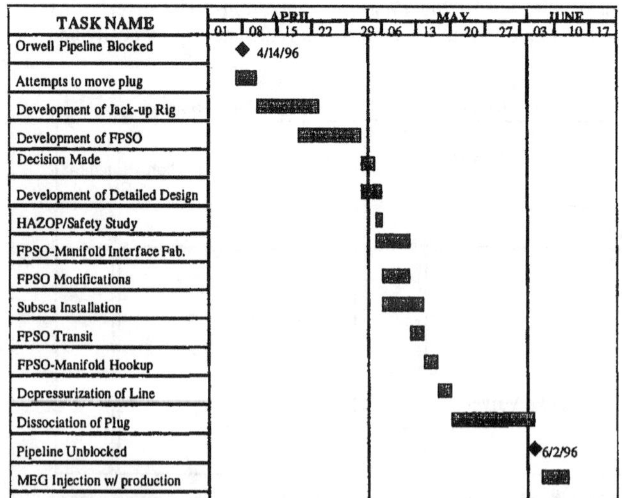

Fig. 3.24—Schedule for complete plug remediation (after Lynch[16]).

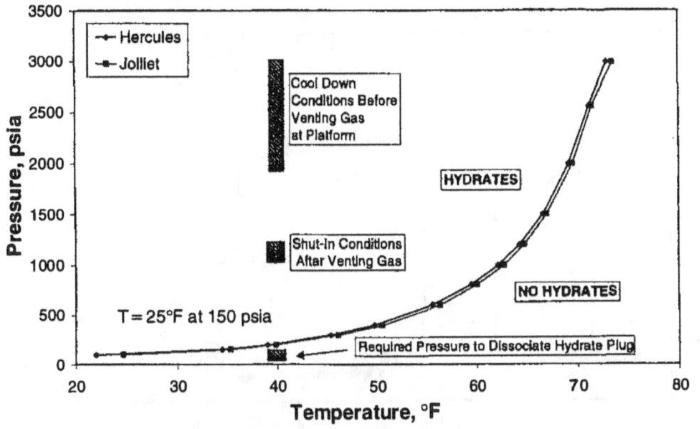

Fig. 3.25—Hydrate-formation conditions (after Ref. 1).

To initiate hydrate dissociation, the hydrostatic head must be removed below 200 psia to approximately 150 psia where the equilibrium temperature is 25°F, slightly inside the ice-formation region, so that a 15°F temperature gradient will cause heat to flow from the ocean to the hydrate.

In a worst-case scenario, the entire volume from the platform to the manifold must be removed. Assuming that only 70% of the pipeline volume is filled with liquid, the volume to be removed would be 12,000 bbl in an 8-in. line and 26,000 bbl in a 12-in. line 50 miles long. The techniques listed in **Table 3.3** were considered for liquid-head removal.

All the options in Table 3.3 require that the plug location be determined and that the pipeline have access points to remove the pressurizing liquid and plug. If there are no access points, the line may be hot-tapped. The figures in the example indicate that workover vessels need to be positioned above the plug.

Of the 7 options summarized in Table 3.3, those with gas lift were eliminated because of low liquid-removal rates. None of the depressurization options were recommended; however, multiple access ports at 4-mile intervals were recommended with use of coiled tubing as described in Sec. 3.4.4 on mechanical removal.

An alternative to pumping the fluids to the surface is to discharge the fluids into a parallel, unblocked flowline. This method requires access points along the pipeline to locate the plug and remove the liquids to the parallel pipeline.

Depressurizing One Side of Plug(s). Rule-of-Thumb 21 indicates that multiple hydrate plugs should be assumed to exist in a shut-in line. With multiple plugs, substantial gas may be trapped between the plugs and depressurization techniques should be similar to depressurization through one side of a plug. The overriding safety concern is that a plug might dislodge from the pipe wall to become a projectile that ruptures a line or vessel.

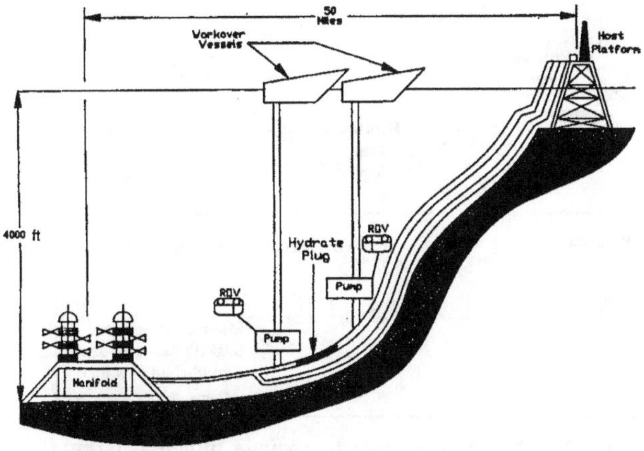

Fig. 3.26—Pipeline depressurization methods—multiphase-pump option (after Ref. 1).

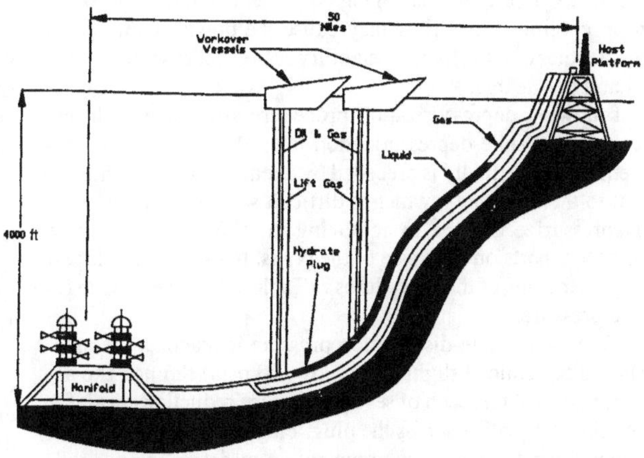

Fig. 3.27—Pipeline depressurization through gas lift (after Ref. 1).

The following procedure may be used for depressurizing one side of a hydrate plug. A similar procedure can be used with multiple hydrate plugs when liquid heads exist on each side of the plug. **Fig. 3.28** illustrates the situation for removal of two hydrate plugs without an intermediate access point.[1] In this case, it is assumed that there are multiple access points to the pipeline so that the general position of the plug(s) can be located by pressure differential. The procedure listed next was slightly modified from that proposed in Refs. 1 and 17.

Procedure for Depressurization of One Side of a Hydrate Plug (Also for Multiple Plugs Without an Intermediate Access Port). When the only option is to depressurize one side of a hydrate plug, there are two major concerns for plug removal: (1) that the plug may

TABLE 3.3—TECHNIQUES TO REMOVE LIQUID HEAD ABOVE A DEEPWATER HYDRATE PLUG

Option for Removing Liquids	Issues/Limitations
1. Multiphase pumping to surface (**Fig. 3.26**) at a rate of 5,000 BOPD to remove liquids in 3 to 6 days	Temporary deployment; electrical submersible pump; handle large liquid volume on workover vessel
2. Subsea separator; vent gas and pump liquid to surface	Deploy separator/pump hardware subsea
3. Gas-lift pipes on each side of plug (**Fig. 3.27**)	Extremely slow: 21 days to remove 12,000 bbl from 8-in. line; 25+ days to remove 26,000 bbl from 12-in. line
4. Multiphase pumping with gas lift	Similar issues to Option 1
5. Combine subsea separator with gas lift	Too slow; similar issues to Options 2 and 3
6. Displace with nitrogen from platform	Requires large volumes of N_2 at high pressure
7. Launch a gel or foam pig followed by nitrogen	Gel pigs separate gas and liquid; access point must be large enough to introduce pig

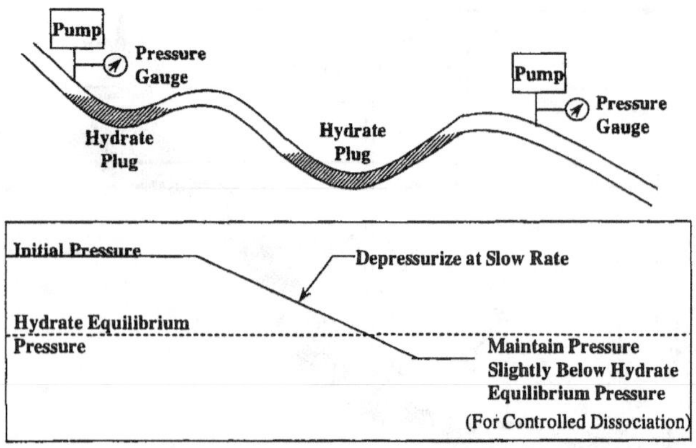

Fig. 3.28—Suggested procedure to remove multiple hydrate plugs (after Ref. 1).

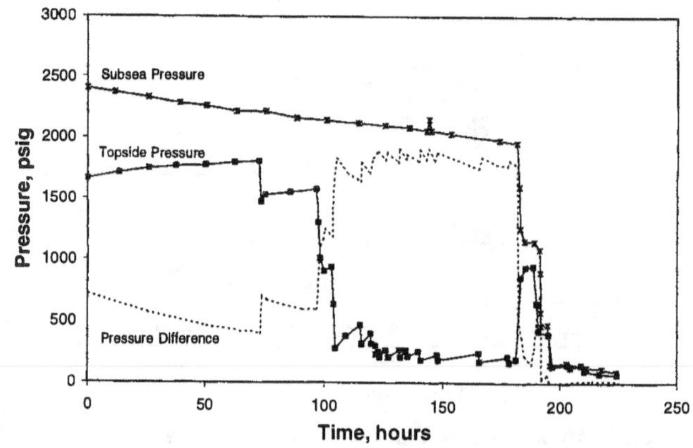

Fig. 3.29—Pressure change during depressurization (after Gjertsen[12]).

dislodge and be propelled in the pipe, becoming a severe safety problem (see Chap. 1) and damaging equipment, and (2) that Joule-Thomson cooling of gas flow may cause the downstream end to progress farther into the hydrate-stability region because the plug is porous and permeable.

The following depressurization procedure attempts to address both concerns. While depressurization is used most often for hydrate removal, it normally is preceded by attempts to place inhibitor adjacent to the blockage—which is difficult without fluid flow.

1. Depressurize the line by removing the fluids at a slow rate though access ports on each side of the plugs. If a substantial liquid head is present, one of the procedures in Table 3.3 can be used to reduce the pressure.

2. Before the hydrate-dissociation pressure is reached, the pressure should be reduced slightly (e.g., by 100 psia) through the access-port valves. After each of several pressure reductions, wait for the pressure to equalize across the plug. Plug permeability and porosity may permit pressure communication to determine gas volumes on each side. While the hydrate plugs are porous, as indicated in the Gullfaks case, pressure equalization may be as slow as 3 psi/hr if substantial liquid flows through the plug.

3. Maintain a low Δp across hydrate plugs; this reduces the threat of a projectile by providing both a low driving force and a downstream gas cushion (see Example 3.3) for any dislodged plug. In addition, a low Δp across the plug minimizes Joule-Thomson cooling at the plug discharge end.

4. Reduce the pressure in stages to a level slightly below the equilibrium pressure, pausing for equilibration at each stage.

5. Monitor the line pressure. If hydrates are dissociating (but remain in the line) the pressure will rise slowly to a level equal to the hydrate-equilibrium pressure at the ocean-bottom temperature. If hydrates have dissociated, the line pressure will remain below the hydrate-equilibrium pressure.

6. Check for Δp across the section. When the plug completely dissociates, there will be no Δp across the section that had contained the plug and Sec. 3.5 should be consulted for system startup.

While this method represents an ideal depressurization from only one side, frequently a nonideal depressurization must be achieved, as in the following case study for a plug with low liquid permeability and a very low GOR. It should be noted that liquid permeability through a hydrate plug is a factor of approximately 1,000 lower than that of gas.

Case Study 3.7. Line Depressured From One Side for Hydrate-Plug Removal. In January 1996, Statoil depressured a hydrate plug in a North Sea line that was used alternatively as a black-oil producer and a gas injector to maintain reservoir pressure.[12] The oil and water production rates were 18 Mscf/D and 2,892 B/D, respectively, and the GOR was usually 100 to 360 scf/scf—a fairly low value. Case Study 3.4 in Sec. 3.3.2 documents the plug-location method.

Because the plug was about midway along the 1.6-mile pipeline and pipeline topology prevented inhibitor contact with the plug, inhibitor melting was not an option. Because there were no connections at the well, the plug had to be depressurized from the platform side only. Consideration of the hydrate-formation curve showed that the plug-equilibrium pressure was 261 psia but that ice would form when the pressure was below 115 psia.

Fig. 3.29 shows the depressurization of the line, with the upstream pressure, the platform pressure, and the pressure drop. During dissociation, the pressure was decreased in steps and a slow bleed through was observed from 0 to 73 hours, from 73 to 90 hours, from 95 to 105 hours, and from 105 through 120 hours.

During the time before 120 hours, the pressure was above the hydrate-equilibrium pressure; and, while the upstream pressure decreased steadily, it never decreased to the downstream pressure, indicating that the plug was not very permeable to black oil. A second possibility was that the light oil ends may have been flashing to maintain a constant pressure upstream. However, the increase in downstream pressure occurred much more rapidly as the downstream pressure was lowered, indicating that the plug was porous, even to the black oil.

After approximately 120 hours, the line pressure was maintained between 145 and 261 psia downstream of the plug. The plug dissociated approximately 50 to 60 hours after the downstream pressure had been reduced sufficiently for melting by heat influx from the ocean, as indicated in Fig. 3.29, by a sudden upstream-pressure decrease from 1,890 to 1,160 psig, while the downstream pressure increased from 218 to 1,015 psig during the same period. The pressure was decreased to 145 psig and kept there for more than 30 hours to melt the remainder of the hydrates.

Restart of the well (see Case Study 3.10) was accomplished 2 weeks after the original plug developed. This case is another indication of the long times required to remediate a hydrate plug.

Case Studies C.25 through C.27 in Appendix C are an overview of DeepStar Wyoming (onshore) field studies of hydrate formation and of dissociation from one side of the plug. These studies had the best instrumentation of any hydrate studies to date and provide several exceptions to the concepts in this portion of the handbook. For example, in two of three cases, relatively impermeable plugs were formed, one of which withstood a Δp of 475 psi and was propelled down the pipeline at a velocity of 270 ft/sec.

In each DeepStar field trial, depressurization was done gradually in stages from one side of a hydrate plug with prior testing to ensure that an absorbing gas cushion existed downstream. Where the hydrate plug existed upstream of an above-ground bend, angle, or valve, the test was aborted and the plug was depressured from both sides for safety reasons.

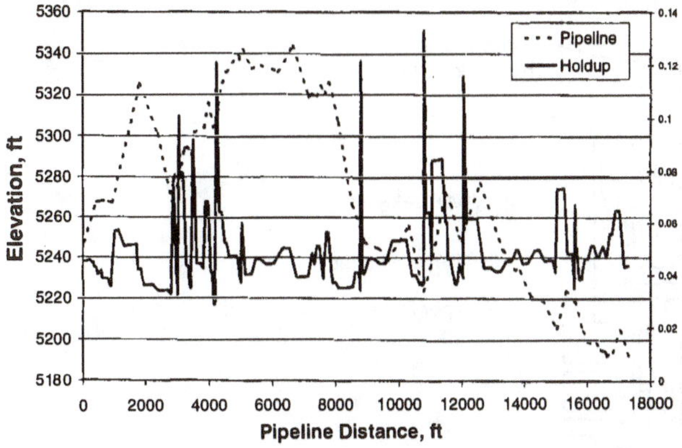

Fig. 3.30—Topography of steady-state holdup profile [after J.J.Xiao and G.Shoup, Amoco letter to G. Hatton, Southwest Research Inst., "OLGA Simulation—Steady-State Results," (16 September 1996)].

In depressuring one side of a hydrate plug, it is instructive to simulate the worst case as a dislodged, frictionless piston projectile in a pipeline, as in Example 3.3.

Example 3.3: Simulation of Hydrate Projectile On Depressuring One Side of Plug. Xiao and Shoup* performed a series of simulations of a hydrate projectile in preparation for depressurization from one side of a hydrate plug in a 4-in. line in Wyoming. The plug was conservatively modeled as a frictionless piston.

With the OLGA® program, the steady-state flow in the line was modeled before blockage formation. The model included pipeline topography to obtain steady-state liquid volumes trapped at low points in the pipeline. The total mass flow was 92 BOPD and 4.166 MMscf/D. **Fig. 3.30** shows pipeline topography and the liquid holdup. At a ground temperature of 34°F, the pipeline was simulated as shut-in for 8 hours, resulting in a plug formation.

Hydrate plugs initially were situated at 7,550 ft from the inlet of a 17,000-ft-long pipeline, with upstream pressures of 1,150 and 575 psig and a constant initial downstream pressure of 50 psig. Transient velocities of two plugs were simulated after formation: (1) a 5-ft-long, 20-lbm plug and (2) a 30-ft-long, 137-lbm plug. Velocity profiles were obtained for both plugs, which were propelled by the initial pressure differentials of 1,100 and 525 psi vs. an initial pressure of 50 psig with a closed valve at the line end.

For upstream pressure of 1,150 psig, the plugs reached peak velocities of 740 ft/sec (smaller plug) and 450 ft/sec (larger plug). For an upstream pressure of 575 psig, the plugs reached peak velocities of 550 ft/sec (smaller plug) and 340 ft/sec (larger plug). The inertial effects of the gas caused rapid acceleration, and the final position of the larger plug (700 and 1,700 ft from the pipe discharge at initial upstream pressures of 1,150 and 575 psig, respectively) was governed by a pressure balance, caused by expansion of the upstream gas and compression of the downstream gas.

The simulation indicated that liquid condensate present in the line had very little effect on the plug maximum velocity when condensate was injected far away from the initial plug position.** **Fig. 3.31** shows the plug velocity as a function of pipe position for the case of 1,150-psig upstream pressure with a 137-lbm plug. Plug simulation results were used to plan and execute field plug-dissociation tests. The calculated plug velocity was an acceptable match with measured plug velocities in the field with a gamma ray detector.

*Xiao, J.J. and Shoup, G.: Amoco letters to G. Hatton, Southwest Research Inst.: "OLGA Simulations—Steady-State Results," (16 September 1996); "OLGA Simulation—Plug Movement Results," (23 September 1996); "OLGA Simulation—Plug Movement Results: Effect of Injecting Condensate on Plug Movement," (6 November 1996); "Flow Pattern Maps for the Kerr-McGee 4 Inch Pipeline," (5 February 1997).

**Xiao, J.J. and Shoup, G.: Amoco letter to G. Hatton, Southwest Research Inst.: "OLGA Simulation—Plug Movement Results: Effect of Injecting Condensate on Plug Movement," (6 November 1996).

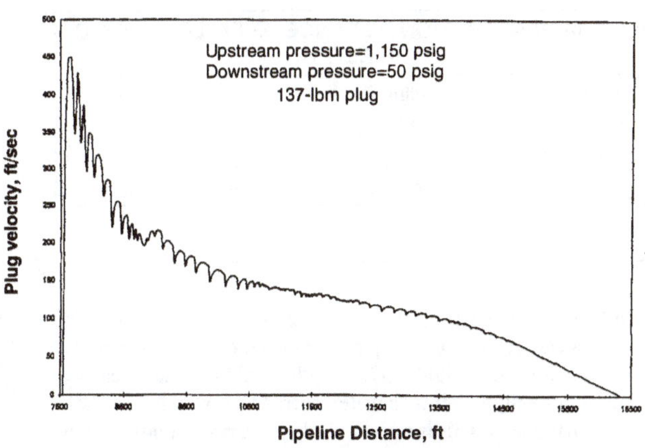

Fig. 3.31—Plug velocity vs. plug location [after J.J.Xiao and G.Shoup, Amoco letter to G. Hatton, Southwest Research Inst., "OLGA Simulation—Steady-State Results," (16 September 1996)].

Note that modeling the plug as a frictionless piston provides conservative results. The modeled plug is slowed by any friction between plug and the pipe as well as by blow-by of gas at the wall and through the porous plug.

3.4.2 Chemical Methods of Plug Removal. When the pipeline is completely blocked, it is difficult to get an inhibitor, such as methanol or ethylene glycol, next to the plug without an access port near the plug. While plugs have proved to be very porous and permeable, particularly in gas systems (see Sec. 3.4.1), a substantial gas volume between the plug and injection points (platform or wellhead) hinders contact, particularly when the line cannot be depressured to encourage gas flow through the plug.

Without flow, inhibitors must displace other line fluids through density differences to reach plugs close to the end of the pipeline. Because flowlines have large variations in elevation, it is unlikely that an inhibitor will reach a plug without flow. Nevertheless, standard practice is to inject inhibitor from both sides of a plug in an attempt to get the inhibitor next to a plug. Sometimes, the increased density of heavy brines can provide a driving force to the hydrate-plug face.

Methanol or glycol injection normally is attempted first in a line. Density differences act as a driving force to get inhibitor to the face of the plug, causing glycol to be used more than methanol, even with glycol's higher viscosity (see Sec. 3.3.2).

3.4.3 Thermal Methods of Plug Removal. When the ends of a hydrate plug cannot be located, heating is very dangerous because the pressure rises exponentially with temperature. Both ends of an undissociated plug can seal the high gas pressure on heating hydrates, and the line can burst as a result. Such a problem is indicated in Case Study 1.4 in Chap. 1.

Rule-of-Thumb 22. Because the ends of a hydrate plug cannot be located easily subsea, nonuniform heating is not recommended for subsea hydrate dissociation.

Heating, however, is a viable option for topside hydrate plugs in exposed pipelines where a thermocamera can be used to determine the plug limits (and where the possibility of multiple plugs has been eliminated). Similarly, in a plugged well where the upper plug end is available, heating may be one of the primary options, as indicated in the next case studies. Heating a plug in a well can be accomplished with a heated wireline broach, similar to the tool shown in Fig. 3.9.

TABLE 3.4—PENETRATION DISTANCE OF COILED TUBING (AFTER REF. 1).

Tubing Size (in.)	Flowline Size (in.)	Penetration (ft)
1.5	4 or 6	3,000 to 5,000
1.75 to 2.0	4 or 6	6,000 to 8,000

Case Study 3.8. Plug Dissociation in a Well. A hydrate plug was experienced in a well feeding a jackup platform in the Norwegian sector of the North Sea in mid-May 1997. A hydrate plug, initially caused by pressurization of the well with water, formed below the downhole safety valve in the well. This is a particularly precarious condition that can result in a well blowout if it is not handled properly.

Field personnel first attempted to decrease the pressure in steps to just above the hydrate-equilibrium pressure, and unsuccessful attempts were made to push MEG through the hydrate plug. The next action was to inject MEG into the well, leaving only a small gas volume at the top of the well. With a higher pressure on top of the plug, the only way to get gas into the well was by hydrate dissociation by MEG.

When the pressure dropped to 4,280 psia, MEG was reinjected into the well until the pressure rose to 4,930 psia. A total of 0.14 gal of MEG was reinjected, indicating that a very small amount of hydrates had dissociated. It was concluded the plug had very low permeability and dissociated very slowly. This ended the period of "getting to know the plug."

At that point, the pressure at the top of the well was reduced to 15 psia and the well was shut in so that only the additional static head (394 ft above the plug) maintained pressure above the plug. The pressure recovered to 100 psia, indicating that hydrates were dissociating on pressure reduction. There were at least six similar pressure-reduction and -recovery confirmations that hydrates were dissociating in the well; each time the pressure increase exponentially approached an asymptote of 100 psia.

The decision was made to keep the pressure at 15 psia on top of the well to provide constant hydrate melting. The plug temperature was approximately 48°F. Five hours after maintaining the pressure at 15 psia, hydrate dissociation was complete and the pressure at the top of the well rose to 160 psia. The entire hydrate plug melted 12 days after the initial formation. Questions remained concerning why the plug did not respond to MEG injection so that depressurization had to be used.

3.4.4 Mechanical Methods of Plug Removal. The traditional types of pigs are not recommended to remove a hydrate plug because compression usually compounds a plug problem. Even for partial plugs, hydrate formation at low-lying points of the flowline may cause the pig to become stuck. If a number of hydrate particles are present in the line, pigging could result in a more severe plug.

Coiled tubing is therefore the final option for offshore hydrate removal. The tubing is inserted through a lubricator, usually at a platform or floating workover vessel, in an effort to get an inhibitor, such as glycol, to the face of the plug.

Coiled tubing of ½- to 3½-in. outer diameter has a maximum length between 15,000 and 29,000 ft.[18] The bend radius at the base of an offshore-platform riser limits coiled-tubing penetration to a minimum radius of 6 to 10 ft, but a preferred minimum radius is 20 to 60 ft. Penetration distance is a function of tubing size and pipeline diameter (**Table 3.4**).

See Case Study 3.3 (Sec. 3.3.2) for a successful example of hydrate-plug removal with coiled tubing and glycol jetting. In other case histories, coiled tubing has been used successfully. For example, coiled tubing recently was used to dissociate a plug at the Statfjord field.* Coiled tubing is expensive, requiring special rigs. The cost to rent the rig for coiled-tubing use in 1997 was U.S. $1 million/D.

*Personal communication with O. Urdahl, Statoil Research Center, Trondheim, Norway (14 July 1997).

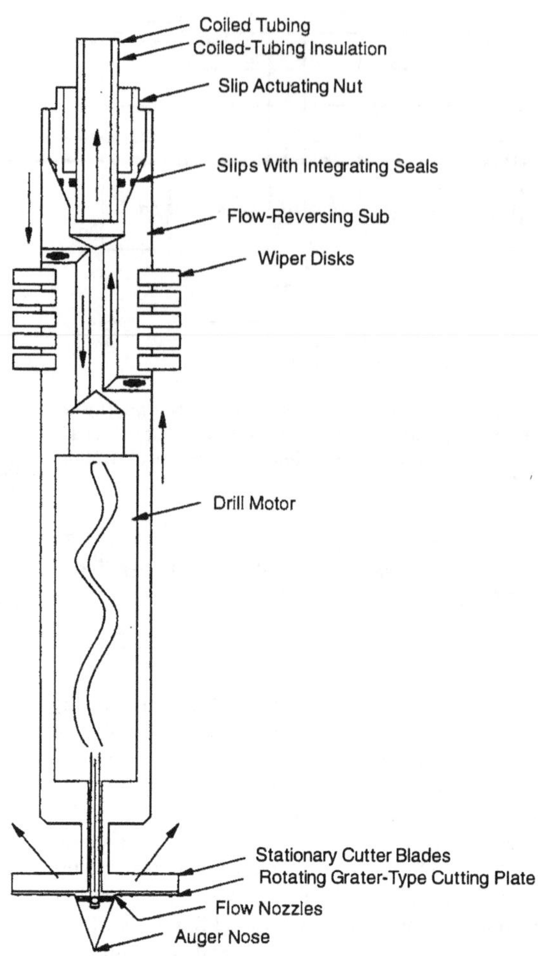

Fig. 3.32—Drilling head for solids removal (after Ref. 1).

Coiled-tubing technology is being improved. For hydrate applications, Ref. 1 lists three new types of coiled tubing:

1. Coiled tubing can get hydraulic drilling equipment to the plug (**Fig. 3.32**).
2. A tractor can be used to pull the coiled tubing through the flowline from the platform side (**Fig. 3.33**) in lines larger than 4-in. ID at a speed up to 5,400 ft/hr with penetration distances to 15,000 ft. Testing was done in Deepstar Project 3202. Also pig "skates" are being developed to improve penetration.
3. Composite coiled tubing is a promising development. The tubing walls are porous to allow air/gas to lubricate the tubing travel for greater penetration. This has not been demonstrated yet.

With the use of coiled tubing, it is important to remember that as much as 180 scf of gas evolves from each cubic foot of dissociated hydrate. Coiled tubing must have gas-flow-by capability in the drive mechanism at the tubing front. This prevents either pushing the tubing from the plug face or line overpressure. For example, with the pig-driven coiled tubing shown in Fig. 3.32, gas must be produced from the tubing.

3.5 Avoiding Hydrates on Flowline Shut-In or Startup

Shut-in and startup are primary times when hydrates form. On shut-in, the line temperature cools very rapidly to that of the ocean floor (40°F for depths greater than 2,000 ft) so that the system is almost always in the hydrate region if the line is not depressured. At that condition, multiple hydrate plugs can form. For a planned shut-in, two actions are recommended: (1) inject a large amount of inhibitor such as methanol or ethylene glycol and (2) depressure the pipeline as soon as possible.

Case Study 3.3 illustrates hydrate-plug formation caused by an unexpected shut-in when methanol could not be injected. It is not clear that the line was depressured immediately after shut-in, but the plug

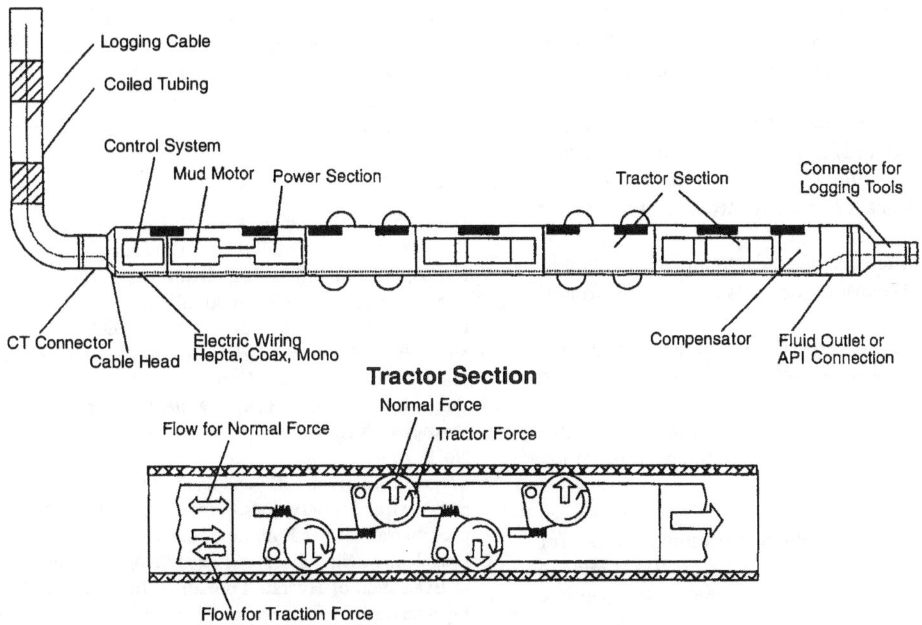

Fig. 3.33—Coiled-tubing tractor (fluid-driven version). (After Ref. 1.)

formation was removed by coiled tubing with glycol jetting. Case Study 3.9 also illustrates the value of line depressuring on shut-in.

Case Study 3.9. Multiple-Plug Formation After Pressurized Shut-In. The following study is from Ref. 1. Because of a problem at a gas plant, a 6-in. 600 ANSI flowline was shut in at 1,000 psig but was not depressured for 6 days. The normal flow in the pipeline was a gas with 2% H_2S and a condensate in the amount of 50 bbl/MMscf.

To remove the blockage, the wellhead side of the line was depressured by venting over a 15- to 20-minute period before the downstream side was vented. During the downstream venting operation, one of the hydrate plugs partially melted, dislodged from the line, and was propelled by the high-pressure gas trapped inside the line. In this case, there were at least two low spots in the line where sufficient water accumulated to form multiple hydrate plugs.

One plug, estimated to be 33 ft long, was propelled rapidly down the pipeline and ruptured a tee near the header within ½ a second after the header valve was opened. The impact of the plug and associated flying debris caused one fatality and one injury to personnel operating the valve. Follow-up investigations and mathematical modeling showed that 230 to 820 ft of high-pressure gas in a 6-in. line would be sufficient to cause the damage that occurred. In subsequent operations, hydrate plugging was prevented by (1) methanol or glycol was injected during each startup, (2) a hydrate inhibitor was injected before stopping flow followed by depressurization for planned shutdowns, and (3) the pipeline was depressured soon after shut-in for unplanned shutdowns.

On startup before reaching steady state, all parts of the system are particularly susceptible to hydrates while the system is heating with warm fluids from the reservoir. During this time, small hydrate particles that have formed may be compacted by flow (or by pigs) to form a plug. A typical startup procedure involves injecting large amounts of inhibitor and using diesel fuel.

Case Study 3.10. Successful Pipeline Startup After Hydrate Formation. In 1996, a black-oil pipeline plug occurred in the Norwegian sector of the North Sea (described in Case Study 3.7). After several precautions, the pipeline was depressured from one side of the plug. When the plug had melted, the line was maintained at atmospheric pressure for more than 1 day to eliminate the light components that might form hydrates.

Before startup, methanol was injected in the amount of 530 gal in the 6-in.-ID, 1.6-mile line from the platform. The pipeline then was pressurized with a large amount of diesel oil from the platform to the subsea valve, indicating that the pipeline was nearly empty of liquid after the previous depressurization to atmospheric conditions. Further injection of diesel corresponding to two pipeline volumes was pumped into the pipeline and well. Subsequently, the well and the pipeline were put into production without any hydrate problems.

3.6 Recommendations and Future Development Areas

3.6.1 Recommendation Summary for Hydrate Remediation.

1. Hydrate plugs are always dissociated, but the time scale usually is days to weeks. Deliberate changes and patience are required. Hourly changes are ineffectual.

2. Multiple hydrate plugs always should be assumed to be present and should be treated as a safety hazard.

3. Many hydrate plugs are porous and transmit pressure easily while acting to obstruct flow. Some plugs are permeable to gas but less so to condensate or black oil. This concept controls many aspects of hydrate dissociation, including radial depressurization, Joule-Thomson cooling through the plug, and the fact that depressurization may cause the plug downstream temperature to decrease below the hydrate-equilibrium temperature. Plug permeability decreases with time.

4. Methods are not well-defined for locating subsea hydrate plugs and determining their length; usually, backpressure and pressure drops are used with some inaccuracy. However, onshore plug location and length may be determined with, for example, gamma ray detection, resulting in more confidence on dissociation.

5. Attempts to blow the plug out of the line with a high upstream pressure always result in a larger, more compacted hydrate.

6. Depressurization from both sides of hydrate plugs is the preferred method of removal, from both safety and technical viewpoints. This implies access points at both plug ends through dual production lines and service lines.

7. If line pressure is decreased rapidly to atmospheric pressure, the hydrate plug will form an ice plug. Surprisingly, this removes the pipeline obstruction more rapidly.

8. With significant elevation change, a liquid head on a hydrate plug may be sufficient to prevent depressurization. Liquid-head removal around subsea hydrate plugs is a challenge to flow assurance.

9. One-sided depressurization of a plug can be done safely but is time consuming because the line must be depressurized very slowly.

10. Heating is not recommended for hydrate plugs without a means of relieving the excess gas pressure when hydrates dissociate.

11. Coiled tubing represents the primary mechanical means for dissociating hydrates.

12. Methanol or glycol usually is injected into plugged flowlines, but this is seldom effective because of the necessity to get the inhibitor at the face of the plug.

13. Inhibitor injection and depressuring techniques are suitable techniques for system shut-in and startup—two times of jeopardy in formation of hydrate plugs.

14. After hydrate plugs are dissociated and substantial inhibitor has been injected, the residual water should be removed promptly (for example, with a pig).

3.6.2 Recommendations for Future Work. Recommendations for future work to aid remediation supplements are from Ref. 1 and are based on case studies represented in the body of this book and in Appendix C.

1. Investigate the use of various access points along a flowline to allow (1) location of the plug, (2) removal of liquid head at each side of a plug, and (3) depressurization from each side of the plug. Such options include multiple access points along a pipeline, dual production lines, wellhead access through service lines with check valves removed or bypassed, and blind flanges and valves at a manifold.

2. Investigate the use of various coiled-tubing techniques to enter a long-distance subsea line, such as a locomotive-type device for pulling coiled tubing, pigs mounted outside of coiled tubing to assist penetration, and composite coiled tubing to reduce drag.

3. Consider using a long-radius riser (from 20 to 80 ft) where possible. Eliminate bends and "S" configurations where water might accumulate. Reduce line low spots.

4. Eliminate unnecessary restrictions and valves in the system and provide for heating or methanol injection where Joule-Thomson cooling is a problem. Consider installing a heater to prevent hydrate formation in the choke and/or separator.

5. Consider providing pressure and temperature monitors a various points along the pipeline. Provide for hydrate prevention at these instrument points.

6. Refine and verify a mathematical model to include radial dissociation of a hydrate plug. A proven, predictive model for hydrate dissociation is not available.

Nomenclature

A = pipeline cross-sectional area, L^2
D = inside diameter of pipeline, L, in.
H = hydrate
ΔH = enthalpy change, mL^2/t^2
I = ice phase
L_w = liquid water
n = value obtained from gas input
p = minimum wellbore pressure, m/Lt^2, psi
Δp = change in minimum wellbore or flowline pressure, m/Lt^2, psi
R = universal gas constant, [10.73(psia-ft^3)/(lbmol-°R)], lbf·L/lbm mole °R
t = time, t
T = inhibitor temperature depression, T, °F
T_H = temperature of hydrate pipe, T, °F
T_I = temperature of ice, T, °F
T_o = ambient temperature outside pipeline, T, °F
T_w = water temperature, T, °F
ΔT = temperature change on heating ($T_2 - T_1$), T, °R
V = volume of downstream plug, L^3
x = radial distance from pipe center, L, in.
X_1 = moving boundary nearest the wall, L, in.
X_2 = moving boundary nearest the pipe center, L, in.
Z = gas compressibility as a function of p and T

References

1. Davalath, J.: "Methods To Clear Blocked Flowlines," report DeepStar IIA CTR A208-1, Mentor Subsea (January 1996).
2. Hatton, G.J., Barajas, A.M., and Kuhl, C.A.: "Hydrate Plug Decomposition Test Program," final report, SWRI Project 04-8217, prepared for DeepStar 3204 Subcommittee (October 1997).
3. Sloan, E.D. Jr.: *Clathrate Hydrates of Natural Gases*, second edition, Marcel Dekker Inc., New York City (1998).
4. Berge, L., Gjertsen, L.H., and Lysne, D.: "Measured Permeability and Porosity of R11 (CCl3F) Hydrate Plugs," *Chem. Eng. Sci.* (1998) **53**, No. 9, 1631.
5. Berge, L., Gjertsen, L.H., and Lysne, D.: "The Importance of Porosity and Permeability of Hydrate Plugs in Pipes," *Proc.*, Second Intl. Conference on Natural Gas Hydrates, Toulouse, France (1996) 533.
6. Lingelem, M.N., Majeed, A.I., and Stange, E.: "Industrial Experience in Evaluation of Hydrate Formation, Inhibition, and Dissociation in Pipeline Design and Operation," *Proc.*, First Intl. Conference on Natural Gas Hydrates, E.D. Sloan Jr., J. Happel, and M. Hnatow (eds.), New York Academy of Sciences, New York City (1994) **715**, 75.
7. "Paraffin and Hydrate Detection Systems," report DeepStar IIA CTR A212-1, Paragon Engineering Services Inc. and Southwest Research Inst., San Antonio, Texas (April 1996).
8. "Pipeline/Flowline Pigging Strategies" report DeepStar II CTR 640-1, tutorial and review, H.O. Mohr Research & Engineering Inc., Houston, Texas (August 1994).
9. Gudmundsson, J.S., Hvedling, F., and Borrehaug, A. "Transporting Natural Gas as Frozen Hydrate," *Proc.*, 5th Intl. Offsite and Polar Eng. Conference, The Hague (1995) 282.
10. Corrigan, A. *et al.*: "Trials of Threshold Hydrate Inhibitors in the Ravenspurn to Cleeton Line," *SPEPF* (November 1996) 250.
11. Austvik, T. *et al.*: "Formation and Removal of Hydrate Plugs, Field Trial at Tommeliten," *Proc.*, Gas Processors Assn. Conference, San Antonio, Texas (1997).
12. Gjertsen, L.H. *et al.*: "Removal of a Gas Hydrate Plug From a Subsea Multiphase Pipeline in the North Sea," *Proc.*, BHR Group 1997 Multiphase Conference, Cannes, France (1997).
13. *GPSA Engineering Data Book*, 10th edition, Gas Processors Suppliers Assn., Tulsa, Oklahoma, 1994.
14. Kelkar, S.K., Selim, M.S., and Sloan, E.D. Jr.: "Hydrate Dissociation Rates in Pipelines," *Proc.*, 13th Symposium on Thermophysical Properties, Boulder, Colorado (1997).
15. Yousif, M.H. *et al.*: "Depressurization of Natural Gas Hydrates in Berea Sandstone Cores," *J. Inclus. Phenom. and Molec. Recog.* (1990) **8**, 71.
16. Lynch, P.: "Orwell Field Pipeline Blockage Report, April–June 1996," report, Arco British Ltd. (November 1996).
17. King, R. *et al.*: *CAPP Guideline for the Prevention and Safe Handling of Hydrates*, Canadian Assn. of Petroleum Producers (January 1994).
18. Sas-Jaworsky, A. II, Blount, C.G., and Tailby, R.J.: "Coiled Tubing Operations and Services," *World Oil* (June 1993) 55.

SI Metric Conversion Factors

°API	141.5/(131.5 + °API)		= g/cm^3
bbl	× 1.589 873	E−01	= m^3
°F	(°F−32)/1.8		= °C
ft	× 3.048*	E−01	= m
ft^2	× 9.290 304*	E−02	= m^2
ft^3	× 2.831 685	E−02	= m^3
gal	× 3.785 412	E−03	= m^3
in.	× 2.54*	E+00	= cm
in.^{2}	× 6.451 6*	E+00	= cm^2
mile	× 1.609 344*	E+00	= km
psi	× 6.894 757	E+00	= kPa
°R	5/9		= K
ton	× 9.071 847	E−01	= Mg

Chapter 4
Economics

4.1 Introduction

Economics provides the motivation for all engineering action. When we ask why hydrates should be of concern, the ultimate answer relates to economics. Even concerns of higher value (e.g., safety or the environment) relate directly to economics because these concerns can prevent process operations.

This chapter provides economics of hydrate safety, prevention, and remediation—the previous three chapters of the book. In every example, a time stamp enables the reader to update the economics with such tools as the *Consumer Price Index* (see the Bureau of Labor Statistics web site at http://stats.bls.gov) or the *Chemical Market Reporter.*[1]

4.2 Economics of Hydrate Safety

Even though insurance actuaries can set a price on life and limb, an ethical concern for worker well-being usually dictates safe operation. Most oil and gas companies take well-deserved pride in their number of accident-free days and often use housekeeping inspections as an oblique means to express safety concerns. Safety is related to cost, but company policy invariably is "Safety at all costs" or "If we cannot operate safely, we cannot operate."

The five case studies in Chap. 1 and Case Study 3.9 indicate a direct relationship between safety and cost because blowout and severe process damage occurred in all cases. Lysne[2] lists three such incidences in which hydrate projectiles erupted from pipelines at elbows and caused the loss of three lives and more than U.S. $7 million in capital costs.

4.3 Economics of Hydrate Prevention

The guidelines for hydrate-prevention design (Sec. 2.11) are certain to involve economics related to individual cases (for example, the cost of a heating system installed around an instrument gas-control valve). Frequently, such costs can be minimized in the original process design without expensive retrofits to correct deficiencies. In this chapter, we discuss the economics of two principal prevention means: chemical injection and heat management.

4.3.1. Chemical Injection Economics. In the U.S. in 1996, the oil and gas production industry used an estimated 400 million lbm of methanol (MeOH), the most-used hydrate inhibitor.* Shell's MeOH usage in deep water is forecast at 50 million lbm/yr. With expanding deepwater work, the use of MeOH is expected to grow 50 to 75% in the next 5 years. These economics provided the initial motivation to investigate hydrate prevention by other means.

*Personal communication with C. Houston, Whitco Inc., Houston (28 July 1997).

Economics of MeOH and Monoethylene Glycol (MEG). Ref. 3 provides of the most comprehensive documented economic studies of MeOH injection. In that work, chemical injection costs (including MeOH) were reported for two U.S. Gulf of Mexico cases: (1) the Jolliet reservoir, which is naturally gas lifted, and (2) the Hercules reservoir, which has a heavier crude and a low gas/oil ratio (GOR) (500 scf/bbl).

The study recommends the following main features.

- One surface pump per chemical on the host platform.
- One subsea transmission line per chemical.
- Subsea distribution with remotely adjustable, pressure-compensated flow-control valves packaged into control pods.
- Use of steel or stainless steel subsea chemical transmission lines.

Table 4.1 provides details of annual hydrate chemical costs for 1 and 20-well cases with 60-mile lines. **Table 4.2** gives capital costs for MeOH injection systems in 1 well and 20 wells for the Jolliet and Hercules reservoirs. It should be noted, however, that both tables are based solely on MeOH only in the free-water phase. As noted in Secs. 2.7.3 and 2.7.4, MeOH losses to the vapor and condensate phases frequently are quite important.

The amounts of chemical injection should be based on the methods of Sec. 2.7, recalling the relative advantages and disadvantages of each inhibitor. For example, MeOH dissolves significantly in the vapor and liquid hydrocarbon phases, not just in the free-water phase. Table 4.1 considers this effect.

MeOH had a delivered cost to an offshore Gulf of Mexico platform of U.S. $2/gal during the 1996–97 winter. Such costs fluctuate significantly and are somewhat seasonal; dockside North Sea MeOH costs were U.S. $0.11/lbm (U.S. $0.72/gal) and ethylene glycol dockside costs were U.S. $0.27/lbm during the summer of 1997.

Because MeOH recovery typically is not economical, MeOH injection normally is considered an operating cost. Ref. 4 shows MeOH recovery to be very expensive in Table 2.11 of Case Study 2.2. For MeOH recovery late in the life of a field, the total installed cost on an existing platform was estimated at U.S. $16.7 million (U.S. $20 million total installed cost with a new platform), while the annual operating cost is U.S. $6 million. For MEG, a low vapor pressure results in a smaller, higher temperature-recovery column, making the economics much more favorable.

Rule-of-Thumb 23. MeOH loss costs can be substantial when the total fraction of either the vapor or the oil/condensate phase is very large relative to the water phase.

Secs. 2.6 and 2.7 provide a quantitative means to validate Rule-of-Thumb 23. Example 2.7 provides a conservative sample calculation

TABLE 4.1—COST OF MeOH USAGE FOR JOLLIET AND HERCULES RESERVOIRS IN GULF OF MEXICO[3]

Reservoir	Number of Wells	Use Life (years)	WHP (psia)	Oil (B/D)	Gas (Mscf/D)	H_2O (B/D)	Subcool ΔT (°F)	MeOH (wt%)	MeOH (gal/min)	Cost (U.S. $1,000/yr)
Jolliet	1	1	3,317	2,500	1,670	2	46.3	35.3	0.026	7.65
		5	1,970	600	3,268	17	38.8	33.1	0.206	60.6
		8	911	43	850	4	27.8	27.3	0.040	11.8
	20	1	2,821	4,400	2,948	4	43.9	34.8	0.051	15.0
		5	1,449	16,400	33,948	124	34.4	31.1	1.412	415.6
		10	1,123	5,100	36,210	172	30.8	29.2	1.832	539.2
Hercules	1	1	2,325	1,367	869	0	41.2	33.9	0	0
		5	1,737	465	376	666	37.0	32.3	7.889	2,322
		8	1,824	23	30	22	37.7	32.6	0.263	77.4
	20	1	2,325	2,700	3,000	0	41.2	33.9	0	0
		5	1,064	22,700	12,500	4,540	30.0	28.7	47.75	14,054
		10	1,064	19,100	11,700	5,157	30.0	28.7	54.24	15,964

WHP = wellhead pressure.

TABLE 4.2—TRANSMISSION LINES (60 MILES) SIZING, COSTS, AND PUMPING SKID COSTS[3]

Reservoir	Number of Wells	Minimum Line ID (in.)	Line Cost (U.S. million $)	Skid Cost (U.S. thousand $)
Jolliet	1	0.306	1.03	5.20
Jolliet	20	0.780	1.11	30.00
Hercules	1	1.629	1.79	34.00
Hercules	20	2.815	1.79	89.50

Additional cost of valve, actuator, manifolding, and packaging = U.S. $6,700/well.

TABLE 4.3—COST PENALTIES FOR MeOH IN PROPANE*

MeOH Concentration in Condensate (ppm by weight)	Reduction in 1993 U.S. Price (U.S. $/metric ton**)	Comment
0 to 30	0	
30 to 50	0 to 2	
50 to 100	2 to 4	Equivalent to $0.25 to $0.50/bbl
100 to 200	4 to 6	Excludes some crackers
200 to 300	6 to 9	Excludes most crackers
>300	9 to 40	Customer reduced confidence

*Personal communication with T. Austvik, Statoil Research Center, Trondheim, Norway (July 1997).
**1 metric ton = 2,205 lbm.

in which 15% of the MeOH is lost to the vapor and liquid hydrocarbon. **Table 4.3** shows a cost penalty for different MeOH concentrations (>30 ppm by weight) in a condensate.

Environmental concerns are also a major driving force to find alternatives to MeOH. MeOH is toxic and should be prevented from entering the water (by dumping overboard) and, subsequently, the food chain. Such concerns currently are very prominent in the Norwegian sector of the North Sea and may take precedence over economics. In the near future, it may be that use of MeOH is less expensive but prohibited by environmental concerns.

Economics of New Types of Inhibitors. Notz[5] provided one of the best comparisons of operating costs for MeOH with kinetic inhibitors in **Tables 4.4 and 4.5** for a field in the North Sea.

Grainger* compared inhibition costs of MeOH, glycol, and a threshold hydrate inhibitor (THI), which consisted of kinetic inhibitors, a corrosion inhibitor, and a solvent. **Table 4.6** represents dock delivery costs, not including shipping to the platform.

*Personal communication with M. Grainger, TR Oil Services, Aberdeen, U.K. (3 September 1997).

From Table 4.6, operating cost benefits of THI appear better than MEG but worse than MeOH. Bloys *et al.*[6] suggested that economics were favorable for new developments (resulting from, for example, the capital savings of avoiding regeneration systems) but marginal for retrofits of systems with traditional inhibitors such as MEG.

The incentive for newer kinetic control methods is a substantial capital-cost reduction by eliminating the need for offshore platform equipment and a small operating cost reduction. The economics are less certain for onshore processes where space is not at a premium. In one high-water-production North Sea field, BP reckoned the capital costs savings at U.S. $50 million for platform costs, including MeOH injection costs, glycol drying, and regeneration.* For example, BP currently operates some southern North Sea pipelines wet, thereby saving the capital cost of drying the gas on the platform. In addition to capital cost for platform processes, a weight/space savings may be realized on the platform itself.

*Personal communication with C.B. Argo and C.G. Osborne, BP Sunbury Research Centre, U.K. (18 July 1997).

TABLE 4.4—RELATIVE USAGE OF MeOH AND KINETIC INHIBITOR (KI) IN A NORTH SEA FIELD[5]

Pipeline Diameter (in.)	Phase	Use Life (years)	H_2O Average (B/D)	Time in Hydrate Zone (hours)	Maximum T (°F)	MeOH in H_2O (wt%)	MeOH (1,000 lbm)	Active KI (1,000 lbm)
16	Multiphase	0	304	0	No hydrate	0	0	0
		7	287	2.3	11.7	16	20.9	0.409
		15	150	40.9	31.4	33	19.3	NA
8	Liquid	0	346	0	No hydrate	0	0	0
		7	295	7.9	17.5	20	21.5	0.441
		15	118	43.2	19.4	21	8.8	0.170
12	Gas	0	17	8.4	25.5	28	9.7	0.128
		7	10	24.6	30.8	33	5.9	NA
		15	4	72.9	32.0	33	2.4	NA

NA = conditions too severe for KI

TABLE 4.5—COMPARISON OF MeOH AND KI COST IN NORTH SEA[5]

			Years When KI Is Effective				Over Entire 15-Year Life of Reservoir				
									Replacing MeOH With KI Whenever Possible		
Line		Use Life	MeOH		KI		MeOH		KI*	MeOH**	
(in.)	Phase	(years)	Weight	Cost	Weight	Cost	Weight	Cost	Weight	Weight	Total Cost
16	Multiphase	7 to 9	25.9	7.8	0.36	3.2	72.9	22.1	0.36	50.0	18.4
8	liquid	6 to 15	52.5	15.9	1.05	9.3	52.5	15.9	1.05	0	9.3
12	gas	1 to 4	15.6	4.7	0.03	0.26	33.0	10.0	0.03	17.4	5.5

Note: All weights are in millions of pounds and all costs are in U.S. million dollars.
*Includes the cost of MeOH solvent for the KI.
***MeOH cost in those years when a KI cannot be used because $\Delta T > 27°F$.

TABLE 4.6—COMPARISON OF THREE TYPES OF INHIBITOR COSTS IN THE NORTH SEA*

	Chemical		
	MEG	MeOH	THI
Concentration, wt%/bbl H_2O	15	15	0.25
Quantity, lbm/bbl H_2O	61.7	61.7	0.882
Cost, U.S. $/bbl H_2O	16 to 17	6.5 to 7.5	8 to 10

*Personal communication with M. Grainger, TR Oil Services, Aberdeen, U.K. (September 1997).

Rule-of-Thumb 24. The 1997 fixed-leg North Sea platform cost was U.S. $77,000/ton.*

BP would like to use unmanned platforms, but the inhibitor recovery units on some platforms prevents them from doing so. As additional costs, Edwards also estimated the operation of an inhibitor recovery unit at 2 hr/D operator time, and maintenance requires 600 to 700 hr/yr at U.S. $85/hr in 1997.

The economics of antiagglomerants are much less certain than those for kinetic inhibitors. No documented costs of antiagglomerants are published. However, any consideration of antiagglomerant economics should include such factors as emulsion breaking, recovery, and disposal.

4.3.2. Heat-Management Economics. Of the two heat-management techniques (insulation methods and pipeline heating), only the insulation state-of-the-art economics are sufficiently established. However, developments are reducing the cost of such technology.

Economics of Offshore Insulation. The minimum overall coefficient achievable with a nonjacketed system is 0.3 Btu/hr-ft-°F (Ref. 7) and costs are typically U.S. $50 to $300/ft for pipes with diameters between 8 and 12 in.

Rule-of-Thumb 25. To achieve a desired heat-transfer coefficient of 0.3 Btu/hr-ft-°F, a nonjacketed system costs U.S. $1.5 million/mile. Typical costs of insulation by bundled lines are U.S. $1.5 to $2 million/mile.

*Personal communication with A.R. Edwards, BP Dimlington Terminal, U.K. (17 July 1997).

Figs. 2.44 and 2.45 compare the cost of the three types of insulation for water depths of 6,000 ft over 60 miles at oil-production rates of 25,000 and 50,000 B/D, respectively. If an average $U = 0.3$ Btu/hr-ft^2-°F is required with a flowline pressure of 4,000 psia, bundled flow lines are more cost-effective. Sec. 2.9.4 provides technical details and associated economics.

4.4 Economics of Hydrate Remediation

Production stops when hydrate blockages occur. When coupled with the fact that all hydrate-blocked lines and wells have to be recommissioned, the question arises about how lost production should be treated—i.e., as lost or as deferred revenue. There is consensus that shut-in production should be counted as lost revenue for reasons including the following.

1. Deferred production usually is counted at the end of reservoir life so that the time value of money is considered. A dollar today is worth more than a dollar in the future because of a positive net present value of money.

2. Fields frequently are sold over their lifetimes, and deferred cost means lost revenue during the ownership of a field.

3. Contracts specify delivery and penalties for nondelivery of hydrocarbons.

Production losses caused by hydrates are site specific but are enormous when considered collectively. From the hydrate group with the largest worldwide remediation experience, Austvik* indicated the magnitude of the problem, "At any instant in the North Sea, there is probably a hydrate blockage which requires remediation." As one onshore example, despite large quantities of MeOH injection for hydrate prevention, Todd *et al.*[8] report that 66 hydrate blockages occurred in one well and production line during the winter of 1995–96, resulting in production losses of more than U.S. $240,000.

Offshore hydrate remediation techniques are very costly if they are not included explicitly in the initial design. For example, Case Study 3.6 represented a fortunate instance (in April 1996) of having an extra flange available at the manifold for depressurization. In this case, two solutions were technically feasible.

1. Jackup Rig. Tow a jackup rig to the site and attach a high-pressure riser to the manifold's subsea tree. Flare exiting gas by means of the rig flare stack. The estimated cost was U.S. $2 million, and approximately 8 weeks was needed to locate a suitable rig. The hydrate removal time would have been 12 weeks.

2. Floating Production and Storage Vessel (FPSO). Connect an FPSO with a processing plant and flare to the subsea manifold's fourth (spare) flange. Estimated cost was U.S. $1.9 million, and an FPSO was available for immediate use, reducing the required time to 6 to 8 weeks.

Other techniques, such as the use of coiled tubing, were not available at the time. The final cost of depressurizing the pipeline from two sides was almost U.S. $3 million dollars, without production losses. Even with such high costs, the loss of production usually causes time to be the deciding resource during remediation.

During remediation periods, gas supply usually is met by substitution. However, the borrowing capacity typically is limited to five times the daily capacity, so that gas supplies are purchased from the spot market. Typical nondelivery penalty costs are U.S. $50,000/D (after tax) on a gas production unit of 125 MMscf/D. Nondelivery contract pressures may be eased by considering hydrates as a *force majeure* as done in Case Study 3.6, implying that no penalties should be incurred because there was no human error.

*Personal communication with T. Austvik, Statoil Reseach Center, Trondheim, Norway (13–15 July 1997).

Nomenclature

T = temperature, T, °F
ΔT = change in temperature, T, °F
U = overall heat transfer coefficient, Btu/hr-ft^2-°F

References

1. *Chemical Market Reporter*, Schnell Publishing Co. Inc., New York City.
2. Lysne, D.: "An Experimental Study of Hydrate Plug Dissociation by Pressure Reduction," D. Ing. thesis, Norwegian Inst. of Technology, U. of Trondheim, Norway (1995).
3. "Chemical Injection System Study," DeepStar I CTR 240-1 report, Intec Engineering, Houston (December 1992).
4. "Evaluation of the Impact of Methanol Recycle/Recovery Facilities on Offshore Production Systems," DeepStar II CTR 221-1 report, Paragon Engineering Services Inc., Houston (December 1994).
5. Notz, P.K.: *Texaco Technical Memorandum (TM 94-0145) on Relative Inhibitor Costs* (26 July 1994).
6. Bloys, J.B., Lacey, C., and Lynch, P.: "Laboratory Testing and Field Trial of a New Kinetic Hydrate Inhibitor," paper OTC 7772 presented at the 1995 Offshore Technology Conference, Houston, 1–4 May.
7. "Insulated Flowline Options Identification Study," DeepStar IIA CTR A601-a report, INTEC Engineering Co. (August 1995).
8. Todd, J.L. *et al.*: *Reliability Engineering—Gas Freezing and Hydrates*, Texaco, 1996.

SI Metric Conversion Factors

bbl	× 1.589 873	E − 01	= m^3
Btu	× 1.055 056	E + 00	= kJ
ft	× 3.048*	E − 01	= m
°F	(°F–32)/1.8		= °C
gal	× 3.785 412	E − 03	= m^3
in.	× 2.54*	E + 00	= cm
lbm	× 4.535 924	E − 01	= kg
mile	× 1.609 344*	E + 00	= km
psi	× 6.894 757	E + 00	= kPa
ton	× 9.071 847	E − 01	= Mg

*Conversion factor is exact.

Appendix A
Gas Hydrate Structures and Properties and How They Form

A.1 Introduction

The reader may turn to Ref. 1 for a more complete explanation of the following discussion. Two hydrate conference summaries also provide research and applied perspectives of the hydrate community.[2,3]

Gas clathrates are crystalline compounds that occur when water forms a cage-like structure around smaller guest molecules. While they are more commonly called hydrates, a careful distinction should be made between these nonstoichiometric clathrate hydrates of gas and other stoichiometric hydrate compounds that occur, for example, when water combines with various salts.

Gas hydrates of current interest are composed of water and eight molecules: methane, ethane, propane, isobutane, normal butane, nitrogen, carbon dioxide, and hydrogen sulfide. Yet other apolar components between the sizes of argon (3.5Å) and ethylcyclohexane (9Å) can form hydrates. Hydrate formation is a possibility where water exists in the vicinity of such molecules at temperatures above and below 32°F. Hydrate discovery is credited in 1810 to Sir Humphrey Davy. Because of their crystalline, nonflowing nature, hydrates became of interest to the hydrocarbon industry in 1934 when they first were observed blocking pipelines. Hydrates concentrate gases; 1 ft^3 of hydrates may contain 180 scf of gas.

Hydrates normally form in one of the three repeating crystal structures shown in **Fig. A-1**. Structure I (SI), a body-centered cubic structure, forms with small natural-gas molecules found in situ in deep oceans. Structure II (SII), a diamond lattice within a cubic framework, forms when natural gases or oils contain molecules larger than ethane but smaller than pentane. SII hydrates commonly occur in hydrocarbon production and processing conditions as well as in many cases of gas seeps from faults in ocean environments.

The newest hydrate, Structure H (SH), named for its hexagonal framework, has cavities large enough to contain molecules the size of common components of naphtha and gasoline. Some initial physical properties, phase-equilibrium data, and models have been determined for SH, and one instance of in-situ SH in the Gulf of Mexico has been found. Information on SH is in the fledgling stages, and because this structure may not occur commonly in production systems, most of this appendix therefore concerns SI and SII.

A.2. Hydrate Crystal Structures

Table A-1 provides a hydrate structure summary for the three hydrate unit crystals (SI, SII, and SH) shown in Fig. A-1. The crystals' structures are given with reference to the water skeleton, which is composed of a basic building-block cavity that has 12 faces with five sides per face, abbreviated 5^{12}. By linking the vertices of 5^{12} cavities, one obtains SI; linking the faces of 5^{12} cavities results in SII; in SH, a layer of 5^{12} cavities connects a layer of other cavity types.

Spaces between the 5^{12} cavities are larger cavities that contain 12 pentagonal faces and either two, four, or eight hexagonal faces (denoted as $5^{12}6^2$ in SI, $5^{12}6^4$ in SII, or $5^{12}6^8$ in SH). In addition, SH has a cavity with three square, six pentagonal, and three hexagonal faces ($4^35^66^3$). In Fig. A-1, an oxygen atom is located at the vertex of each angle in the cavities; the lines represent hydrogen bonds with which one chemically bonded hydrogen connects to an oxygen on a neighboring water molecule.

Inside each cavity resides a maximum of one of the small guest molecules, typified by the eight guests associated with 46 water molecules in SI ($2[5^{12}] \cdot 6[5^{12}6^2] \cdot 46H_2O$), indicating two guests in the 5^{12} and six guests in the $5^{12}6^2$ cavities of SI. Similar formulas for SII and SH are ($16[5^{12}] \cdot 8[5^{12}6^4] \cdot 136H_2O$) and ($3[5^{12}] \cdot 2[4^35^66^3] \cdot 1[5^{12}6^8] \cdot 34H_2O$), respectively.

SI, a body-centered cubic structure, forms with natural gases containing molecules smaller than propane; consequently, SI hydrates are found in situ in deep oceans with biogenic gases containing mostly methane, carbon dioxide, and hydrogen sulfide. SII, a diamond lattice within a cubic framework, forms when natural gases or oils contain molecules larger than ethane; SII represents hydrates from most natural-gas systems. Finally, SH hydrates must have a small occupant (like methane, nitrogen, or carbon dioxide) for the 5^{12} and $4^35^66^3$ cages, but the molecules in the $5^{12}6^8$ cage can be as large as 0.9Å (e.g., ethylcyclohexane). SH has not been found commonly in natural-gas systems to date.

A.3 Properties Derive From Crystal Structures

A.3.1 Mechanical Properties of Hydrates. As may be calculated with Table A-1, if all the cages of each structure are filled, all three known hydrates have the amazing property of being approximately 85 mol% water and 15 mol% gas. The fact that the water content is so high suggests that the mechanical properties of the three hydrate structures should be similar to those of ice. **Table A-2** shows that this conclusion is true to a first approximation with the exception of

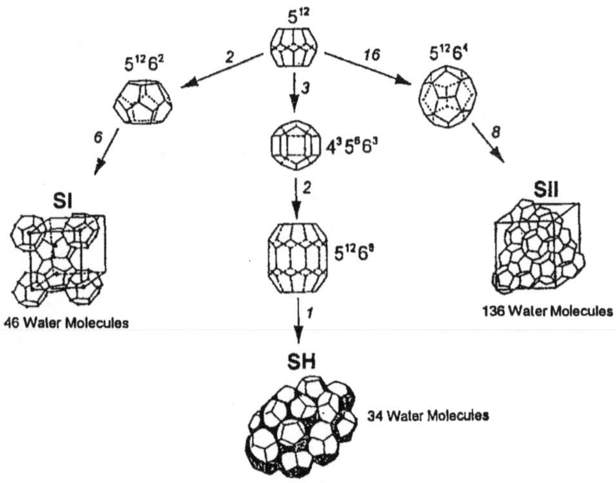

Fig. A-1—Three hydrate unit crystals and constituent cavities.[1]

thermal conductivity and thermal expansivity. Many SH mechanical properties have not been measured.

A.3.2. Guest/Cavity Size Ratio: A Basis for Property Understanding. The hydrate cavity occupied is a function of the size ratio of the guest molecule to the cavity occupied. To a first approximation, the concept of a ball fitting within a ball is a key to understanding many hydrate properties. **Fig. A-2** illustrates five points regarding the guest/cavity size ratio for hydrates formed of a pure guest component in SI or SII.

1. The sizes of stabilizing guest molecules range between 3.5 and 7.5Å. Below 3.5Å, molecules do not stabilize SI, and above 7.5Å, molecules do not stabilize SII.

2. Some molecules are too large to fit the smaller cavities of each structure (e.g., C_2H_6 fits in the $5^{12}6^2$ of SI or *i*-C_4H_{10} fits the $5^{12}6^4$ of SII, but neither fit the 5^{12} cages).

3. Other molecules, such as CH_4 and N_2, are small enough to enter both cavities (5^{12}+$5^{12}6^2$ in SI or 5^{12}+$5^{12}6^4$ in SII) when hydrate is formed of single components.

4. The largest molecules of a gas mixture usually determine the structure formed. For example, because propane and *i*-butane are present in many natural gases, they cause SII to form. In such cases, methane distributes in both cavities of SII and ethane enters only the $5^{12}6^4$ cavity of SII.

5. Molecule sizes that are close to the hatched lines separating cavity sizes exhibit the most nonstoichiometry because of their inability to fit securely within the cavity. For those molecules, nonstoichiometry implies that the water/guest ratio changes most as a function of pressure and temperature.

Table A-3 shows the size ratio of several common gas molecules within each of the four SI and SII cavities. Note that a size ratio

TABLE A-1—GEOMETRY OF CAGES IN THREE HYDRATE CRYSTAL STRUCTURES IN FIG. A-1

	Hydrate Crystal Structure						
	I		II		H		
Cavity	Small	Large	Small	Large	Small	Medium	Large
Description	5^{12}	$5^{12}6^2$	5^{12}	$5^{12}6^4$	5^{12}	$4^35^66^3$	$5^{12}6^8$
Number of cavities/unit cell	2	6	16	8	3	2	1
Average cavity radius, Å	3.95	4.33	3.91	4.73	3.91†	4.06†	5.71†
Variation in radius,* %	3.4	14.4	5.5	1.73	NA	NA	NA
Coordination number**	20	24	20	28	20	20	36
Number of waters/unit cell	46		136		34		

NA = not available.
*Variation in distance of oxygen atoms from center of cage.
**Number of oxygens at the periphery of each cavity.
†Estimates of SH cavities from geometric models

TABLE A-2—COMPARISON OF PROPERTIES OF ICE AND SI AND SII HYDRATES

Property	Ice	SI	SII
Spectroscopic properties*			
Crystallographic unit cell			
Space group	$P6_3/mmc$	Pm3n	Fd3m
Number of H_2O molecules	4	46	136
Lattice parameters	a=4.52 c=7.36	12.0	17.3
Dielectric constant	94	≈58	58
Far infrared spectrum	Peak at 229 cm^{-1}	Peak at 229 cm^{-1} with others	
H_2O diffusion correl time, μsec	220	240	25
H_2O diffusion activation energy, kJ/m	58.1	50	50
Mechanical properties[1]			
Isothermal Young's modulus at 268 K, 10^9 Pa	9.5	8.4**	8.2**
Poisson's ratio	0.33	≈0.33	≈0.33
Bulk modulus	8.8	5.6	NA
Shear modulus	3.9	2.4	NA
Velocity ratio, compression/shear	1.88	1.95	NA
Thermodynamic properties*			
Linear thermal expansion at 200 K, K^{-1}	56×10^{-6}	77×10^{-6}	52×10^{-6}
Adiabatic bulk compression (10^{-11} Pa)	12	14**	14**
Longitudinal speed of sound, km/sec	3.8	3.3	3.6
Transport property*			
Thermal conductivity at 263 K, W/m-K	2.23	0.49 ± 0.02	0.51 ± 0.02

*All properties at 273 K unless noted otherwise.
**Estimated.

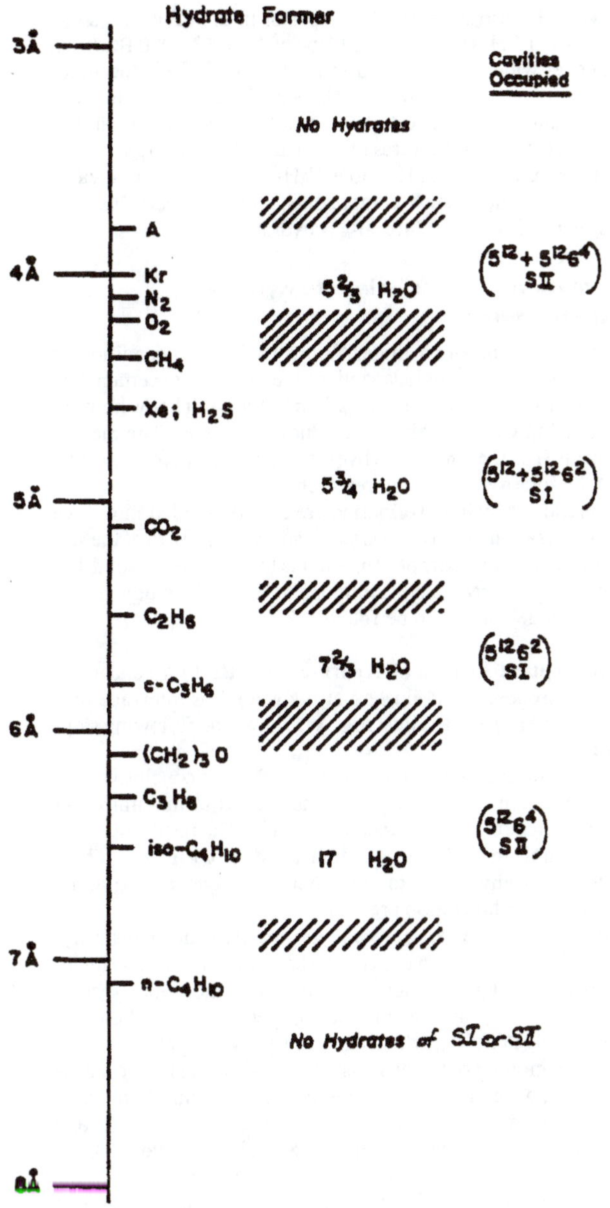

Fig. A-2—Relative sizes of hydrate guest and host cavities.[1]

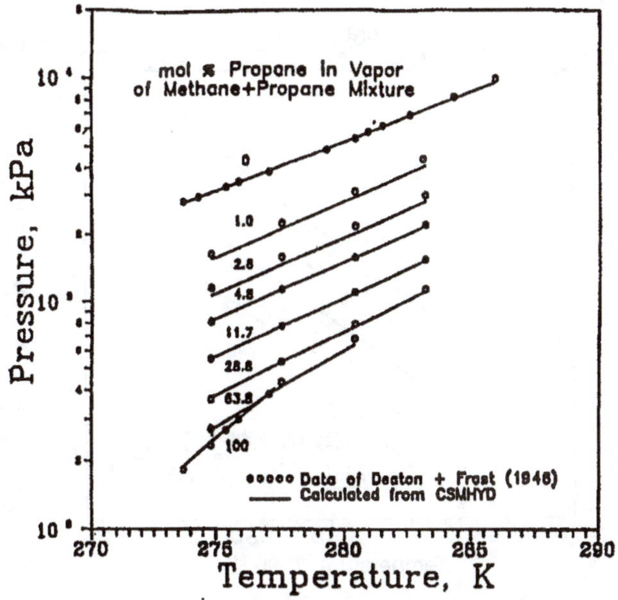

Fig. A-3—Three-phase (L_W-H-V) equilibria of methane/propane mixtures.[1]

(guest molecule/cavity) of approximately 0.9 is necessary for stability of a simple hydrate, indicated by the superscript F. When the size ratio exceeds unity, the molecule will not fit within the cavity and the structure will not form. When the ratio is significantly less than 0.9, the molecule cannot lend significant stability to the cavity.

As Table A.3 shows, ethane as a single gas forms in the $5^{12}6^2$ cavity in SI because ethane is too large for the small 5^{12} cavities in either structure, but too small to give much stability to the large $5^{12}6^4$ cavity in SII. Similarly, propane is too large to fit any cavity except the $5^{12}6^4$ cavity in SII; therefore, gases of pure propane form SII hydrates from free water. On the other hand, methane's size is sufficient to lend stability to the 5^{12} cavity in either SI or SII, with a preference for SI because CH_4 lends slightly higher stability to the $5^{12}6^2$ cavity in SI than to the $5^{12}6^4$ cavity in SII.

A.3.3 Phase Equilibrium Properties. In **Fig. A-3**, pressure is plotted against temperature with gas composition as a parameter for methane plus propane mixtures. Consider a gas of any given composition (marked 0 through 100% propane) on a line in Fig. A-3. At conditions to the right of the line, a gas of that composition exists in equilibrium with liquid water. As the temperature is reduced (or as the pressure is increased), hydrates form from gas and liquid water at the line, so three phases (liquid water plus hydrates plus gas) will be in equilibrium. With further reduction of temperature (or increase in pressure), the fluid phase that is not in excess (water in pipeline environments) will be exhausted, so that to the left of the line the hydrate will exist with the excess phase (gas).

All the conditions given in Fig. A-3 are for temperatures higher than 32°F, and pressures along the lines vary exponentially with temperature. Put explicitly, hydrate stability at the three-phase (L_W-H-V) condition is always much more sensitive to temperature than

TABLE A-3—GUEST-DIAMETER/CAVITY-DIAMETER RATIOS FOR NATURAL-GAS-HYDRATE FORMERS

		(Molecular Diameter/Cavity Diameter)			
		SI Cavity Type		SII Cavity Type	
Molecule	Guest Diameter (Å)	5^{12}	$5^{12}6^2$	5^{12}	$5^{12}6^4$
N_2	4.1	0.804	0.700	0.817^F	0.616^F
CH_4	4.36	0.855^F	0.744^F	0.868	0.655
H_2S	4.58	0.898^F	0.782^F	0.912	0.687
CO_2	5.12	1.00	0.834^F	1.02	0.769
C_2H_6	5.5	1.08	0.939^F	1.10	0.826
C_3H_8	6.28	1.23	1.07	1.25	0.943^F
i-C_4H_{10}	6.5	1.27	1.11	1.29	0.976^F
n-C_4H_{10}	7.1	1.39	1.21	1.41	1.07

F = the cavity occupied by the single hydrate former

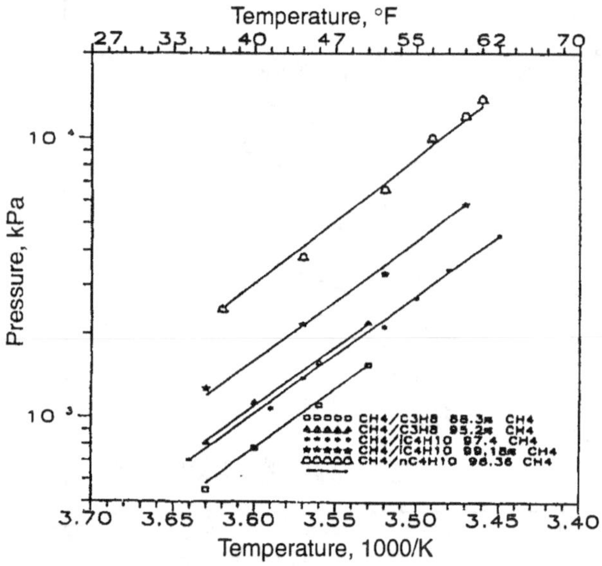

Fig. A-4—Three-phase (L_w-H-V) equilibria of methane/(propane and two butanes).[1]

to pressure. Fig. A-3 also illustrates the dramatic effect of gas composition on hydrate stability; very small amounts of propane added to methane cause structural changes (SI→SII) that form a hydrate with much wider stability conditions. Note that a 50% decrease in pressure is needed to form SII hydrates when as little as 1% propane is in the gas phase.

Any discussion of hydrate dissociation would be incomplete without indicating that hydrates provide the most industrially useful application of statistical thermodynamics prediction of phase equilibria. The van der Waals and Platteeuw model that forms the basis for HYDOFF was formulated after the determination of SI and SII structures shown in Fig. A-1. With the model, one may predict the three-phase pressure or temperature of hydrate formation by knowing the gas composition. For a more detailed discussion, see Chap. 5 of Ref. 1.

A.3.4. Heat of Dissociation. The heat of dissociation, ΔH_d, may be considered to be the heat (rigorously, enthalpy change) required to dissociate hydrates to a vapor and aqueous liquid. For SI and SII, to a fair engineering approximation (± 10%), ΔH_d depends mostly on crystal hydrogen bonds but also on the cavity occupied within a wide range of component sizes.

Enthalpies of dissociation may be determined by the univariant slopes of phase-equilibrium lines (ln p vs. $1/T$) in previous paragraphs, by use of the Clausius-Clapeyron relation [$\Delta H_d = -zR\, d(\ln p)/d(1/T)$]. As one illustration, simple hydrates of C_3H_8 or i-C_4H_{10} have similar ΔH_d of 55,500 and 57,200 Btu/(lbm mol gas) because they both occupy $5^{12}6^4$ cavities, although their guest/cavity size ratios differ (0.943 and 0.976).

As a second illustration, similar slopes of lines in Fig. A-3 show that mixtures of CH_4+C_3H_8 have a value of $\Delta H_d = 34{,}000$ Btu/(lbm mol gas) over wide ranges of composition, wherein C_3H_8 occupies most of the $5^{12}6^4$ cavities, while CH_4 occupies a small number of $5^{12}6^4$ and many 5^{12}. **Fig. A-4** shows similar line slopes (and thus ΔH_d values) for binary mixtures of methane when the large guest is changed from C_3H_8, to i-C_4H_{10}, to n-C_4H_{10}. Because natural gases almost always contain such components, $\Delta H_d = 34{,}000$ Btu/(lbm mol gas) is valid for most natural-gas hydrates.

A.4 Formation Kinetics Relate to Hydrate Crystal Structures

The answers to "What are hydrates?" and "Under what condition do hydrates form?" in the previous sections are much more certain than answers to "How quickly do hydrates form?" We don't know how hydrates form, but we can make some educated guesses. The mechanism and rate (i.e., the kinetics) of hydrate formation are controversial topics at the forefront of current research.

The kinetics of hydrate formation are clearly divided into three parts: (1) nucleation of a critical crystal radius, (2) growth of the solid crystal, and (3) transport of components to the growing solid/liquid interface. All three kinetic components are under study, but an acceptable model has yet to be found.

A.4.1 Conceptual Picture of Hydrate Growth. In a conceptual picture, we proposed that clusters at the water/gas interface may grow to achieve a critical radius; **Fig. A-5** shows the following steps schematically.

1. When natural gases dissolve in water, there is conclusive evidence that water molecules organize themselves to maximize hydrogen bonding around each apolar molecule. The resulting liquid clusters resemble the individual hydrate cavities of Fig. A-1. These fluid clusters are envisioned to join other clusters as the beginning of the hydrate crystallization process.
2. Fig. A-5 indicates an autocatalytic reaction mechanism hypothesized for hydrate formation on the basis of limited experimental evidence. The figure depicts the progress of molecular species from water (Point A), through metastable species (Points B and C), to stable nuclei (Point D), which can grow to large species.
3. At the beginning of the process [species (Point A)], hydrogen-bonded liquid, water, and gas are present in the system. Water clusters around gas molecules to form both large and small clusters (Point B) similar to the hydrate cages of SI and SII. At Point B, the cages are termed labile—they are relatively long-lived but unstable.
4. The cages may either dissipate or grow to hydrate unit cells or agglomerations of unit cells (Point C), thus forming metastable nuclei. Because these metastable unit cells at Point C are of subcritical size, they may either grow or shrink in a stochastic process. The metastable nuclei are in quasi-equilibrium with the liquid-like cages until the nuclei reach a critical radius. After attaining the critical radius (Point D), the crystals grow rapidly in a period sometimes called catastrophic growth.
5. In our conceptual picture, when the system is heated, it is driven to the left in Fig. A-5 and stable hydrate crystals are dissociated.

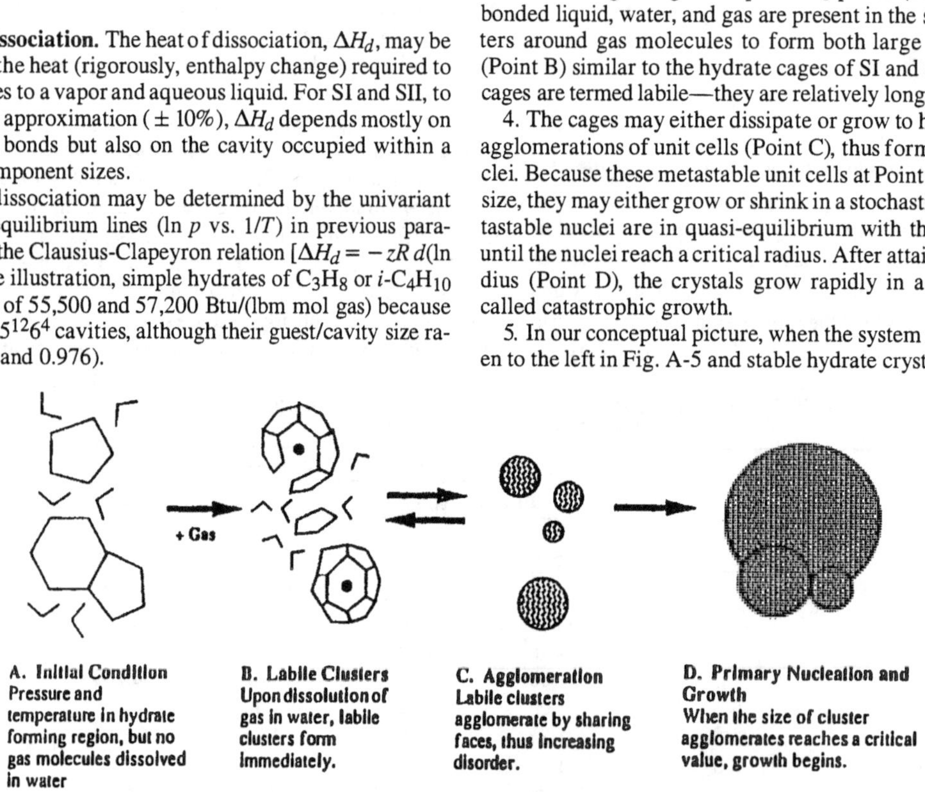

Fig. A-5—Schematic model of hydrate cluster growth.[1]

Once the hydrate-dissociation point is reached and passed, there are still labile microscopic species in the water that range in size from multiple hydrate unit cells (Point C) to metastable nuclei (Point B). These residual structures are present up to a certain level of thermal energy above dissociation. At temperatures below that upper boundary, these species cause a decrease in induction or metastability time of a successive run because the building blocks of crystals remain in the liquid. However, once approximately 100°F is passed, no residual structure remains to promote hydrate formation.

This cluster model conceptual picture is most likely to occur at the interface, either in the liquid or the vapor side. The reader should note that this is a largely unproven hypothesis whose only justification is to serve as a mental picture for qualitative predictions and future corrections.

In contrast to well-determined thermodynamic properties, kinetic characterization of hydrates is very ill-determined. One has only to turn to the review of hydrate kinetics by Englezos[4] or to Ref. 1 to learn the following unsettling facts.

1. Hydrate nucleation is both heterogeneous and stochastic and, therefore, is only approachable by very approximate models. Most hydrate-nucleation models assume homogeneous nucleation and typically cannot fit more than 80% of the data generated in the laboratory of the modeler.
2. Hydrate-growth kinetics are apparatus-dependent; the results from one laboratory are not transferable to another laboratory or field situation.
3. In kinetics and thermodynamics, the hydrate phase is almost never measured directly.
4. The hydrate-dissociation models derived from solid moving-boundary differential equations do not account for the porous, surface formation, and occlusion nature of hydrates on a macroscopic scale.
5. No satisfactory kinetic model currently exists for formation or dissociation. Because of the unsatisfactory state of hydrate kinetics knowledge, this area is the subject of intensive research.

Nomenclature

H = hydrate
ΔH_d = heat of dissociation, Btu/lbm mole gas
L_W = liquid water
p = pressure, m/Lt^2, psia
T = temperature, T, °F
V = vapor
z = gas compressibility, RV/RT

References

1. Sloan, E.D. Jr.: *Clathrate Hydrates of Natural Gases*, second edition, Marcel Dekker Inc., New York (1998).
2. Sloan, E.D. Jr., Happel, J., and Hnatow, M.A.: *Annals of New York Academy of Sciences,* First Intl. Conference on Natural Gas Hydrates (1994) **715,** 519.
3. Monfort, J.P.: *Proc.*, Second Intl. Conference on Natural Gas Hydrates, F. Foucaud, Secretariat, Toulouse Cedex, France (June 1996) 739.
4. Englezos, P.: "Kinetics of Gas Hydrate Formation and Kinetic Inhibition in Offshore Oil and Gas Operations," *Proc.*, Fifth Intl. Offshore & Polar Engineering Conference, The Hague, The Netherlands (1995) 289.

SI Metric Conversion Factors

Å	× 1.0	E − 01 = nm
Btu	× 1.055 056	E + 00 = kJ
°F	(°F−32)/1.8	= °C
°F	(°F+459.67)/1.8	= K
in.	× 2.54*	E + 00 = cm
lbm	× 4.535 924	E − 01 = kg

*Conversion factor is exact.

June 2015
The floppy disk that was originally packaged with this monograph is no longer available. Program files similar to those described in Appendix B can be downloaded from the following website:

http://hydrates.mines.edu/CHR/Software.html

Appendix B
User's Guide for HYDOFF and XPAND Programs

B.1 A Word of Caution

While it is hoped that the programs accompanying this book will be useful in estimating the limiting conditions of hydrate formation, neither the author nor SPE can be held accountable for the use of the predictions that the program provides. If there is a safety consideration or an important process decision to be made on the basis of the program's predictions, the user is cautioned to obtain a second opinion from someone knowledgeable in hydrate-phase equilibria before proceeding.

B.2 Executive Summary

B.2.1 Program Specifications. This program has been developed to run on PC-compatible computers with DOS as the operating system. The program is executable without any additional hardware or software requirements.

B.2.2 Contents of the Disk. The 3.5-in. disk provided with this book contains four files.

1. HYDOFF.EXE,* an executable file to predict hydrate-formation conditions.
2. FEED.DAT, a file to be used as external input of the feed components and composition for HYDOFF. FEED.DAT is an optional file; it should be noted that HYDOFF will run with or without the file FEED.DAT.
3. XPAND.EXE,* an executable file to determine the isenthalpic ($\Delta H = 0$) and isentropic ($\Delta S = 0$) gas-expansion conditions.
4. HYDCALC.XLS,** a short-cut estimation spreadsheet to calculate methanol (MeOH) or monoethylene glycol (MEG) amounts. Sec. 2.6.5 specifies use of this program.

Appendix B provides common examples that use HYDOFF and XPAND; the engineer may modify these for his/her own purposes. Sec. B.3 considers the use of HYDOFF (and FEED.DAT), while Sec. B.4 details the use of XPAND.EXE.

B.3 HYDOFF

B.3.1 Running the Program. The program can be executed directly from the 3.5-in disk or copied to the hard drive and then executed. It is recommended that a backup copy be made of the program in case problems (e.g., virus or erasure) occur. To run the program at the DOS prompt, simply type HYDOFF and follow the instructions given by the program.

B.3.2 Program Overview. The essence of the program is the same as the program included with Ref. 1, to which the reader is referred for a full explanation. The program has the central purpose of predicting hydrate-phase equilibria with and without thermodynamic inhibitors. The version accompanying this book has been abbreviated for rapid use. The program provides pressure predictions of Structure I (sI) and II (sII) hydrates at a given temperature with and without thermodynamic inhibitors [MeOH, salt (NaCl), or mixtures thereof] at three- and four-phase conditions (I-H-V, L_W-H-V, L_W-H-V-L_{HC}).

The method used by the program for hydrate-phase equilibria is based on the van der Waals and Platteeuw model, and the hydrocarbon-fluid phases are modeled with the Soave-Redlich-Kwong equation of state with parameters regressed from experimental measurements.[1]

B.3.3 Specifications for a Problem. Before any calculation is performed by the program, the user is asked to enter some basic information: units in which he/she prefers to operate, components present in the feed, feed composition, temperature, type, and amount of thermodynamic inhibitor(s).

The feed components and composition can be entered directly in the program or specified in the FEED.DAT file, which can be read by the program. Note that the FEED.DAT file must be present in the same directory as HYDOFF.EXE. The units and feed composition can be changed at any point during the execution of the program without actually exiting. When specifying components directly in the program (i.e., not using FEED.DAT for feed input), components can be separated by a space, comma, Enter, or Return.

The program has a MAIN MENU that directs the user to the desired type of calculation. Once a particular calculation is chosen, the user is asked to enter the temperature and, if applicable, concentration of thermodynamic inhibitor(s) in the free aqueous phase.

At no point in the program is the user asked to enter an initial guess for the calculations (for pressure predictions). The program has its own internal initial guess. Also, the user does not have to specify the equilibrium phases for any calculation. The equilibrium phases are given as output of the predictions.

*The HYDOFF and XPAND programs are copyrighted to E.D. Sloan Jr., Center for Hydrate Research, Colorado School of Mines, Golden, Colorado.
**Courtesy of P.J. Shuler, Chevron Petroleum Technology Co.

B.3.4 What To Expect for an Answer. The standard output for hydrate-phase-equilibria calculations displays the following.

1. Equilibrium phases (*I-H-V*, L_W*-H-V*, or L_W*-H-V-*L_{HC}).
2. Equilibrium pressure.
3. Hydrate equilibrium crystal structure (sI or sII).
4. Phase components and compositions (i.e., feed, fluid hydrocarbon, and hydrate).
5. Fractional occupancy of cages by hydrate formers in each type of hydrate cavity.

Different outputs are shown for each calculation type. Examples that follow illustrate how the program is structured and the format of the output.

B.3.5 Some Important Notes. The program is structured to prompt the user whenever incorrect or improper information is entered. The following is a list of limitations and guidelines of which the user should be aware.

1. The maximum number of components is limited to 17.
2. The weight percentage of MeOH as inhibitor is limited to 50 wt%.
3. The freezing-point depression for systems containing both MeOH and salt is determined by additive contributions of MeOH and salt in solution.
4. The total amount of MeOH is assumed to be in the aqueous phase. Possible partitioning of MeOH into other phases (condensate or gas) is neglected. See Sec. 2.7 for calculation details of inhibitor partitioning to other phases.

B.3.6 Examples.

Example B.1: Temperature and Pressure Predictions for Hugoton Gas.[2]

	Component	Mol%
Gas Composition:	Methane	73.29
	Ethane	6.70
	Propane	3.90
	i-butane	0.36
	n-butane	0.55
	Nitrogen	15.00
	n-pentane	0.20
	n-hexane	0.00

Pressure prediction at $T = 51.35°F$

Input Success and Data

HYDRATE PREDICTION PROGRAM. HYDOFF

Press RETURN to continue . . .
AVAILABLE UNITS ARE AS FOLLOWS:

	TEMPERATURE	PRESSURE
(1)	Fahrenheit	psia
(2)	Kelvin	kPa

Please select the desired set of Units:
1
The program has been designed to allow the user to input the feed components and composition directly in the program or through an external file, namely, FEED.DAT.

If the user wishes to read the feed components and composition from FEED.DAT, please make sure the information is entered correctly into FEED.DAT (user has to CHANGE the COMPOSITIONS ONLY) and FEED.DAT is in the same directory as the executable HYDOFF.EXE file.

Are the FEED COMPONENTS and COMPOSITION saved in FEED.DAT (Y or N)?
N
How many COMPONENTS (excluding Water) are present?
8

sI and sII HYDRATE FORMERS

1. Methane	2. Ethane	3. Propane
4. i-Butane	5. n-Butane	6. Hydrogen Sulfide
7. Nitrogen	8. Carbon Dioxide	

NON-HYDRATE FORMERS

9. n-Pentane	10. Isopentane	11. Hexane	12. Heptane
13. Octane	14. Nonane	15. Decane	16. Toluene

Which Components are present? Please list Hydrate formers first.
1 2 3 4 5 7 9 11
Enter the MOLE FRACTIONS of each Component:
Mole Fraction of Methane: 0.7329

Mole Fraction of Ethane: 0.0670

Mole Fraction of Propane: 0.0390

Mole Fraction of i-Butane: 0.0036

Mole Fraction of n-Butane: 0.0055

Mole Fraction of Nitrogen: 0.1500

Mole Fraction of Pentane: 0.0020

Mole Fraction of Hexane: 0.0000

THE FOLLOWING OPTIONS ARE CURRENTLY AVAILABLE
(1) MAIN Program for Equilibrium Hydrate Predictions
(2) Display Feed Composition
(3) Change Feed Composition
(4) Change Program Units
(5) DISCARD all Data and begin NEW Problem
(6) Exit HYDOFF Program
1
PLEASE CHOOSE ONE OF THE FOLLOWING OPTIONS
(1) Pressure prediction at a given Temperature
(2) Pressure prediction at a given Temperature with an inhibitor
(3) Change Feed Composition
(4) Change Program Units
(5) Return to Main Menu
(6) Quit HYDOFF
1
Enter the required Temperature (in F)
51.35

Program Output

THREE-PHASE (Lw-H-V) EQUILIBRIUM CONDITION
Temperature: 51.350 F — **Experimental pressure**
Equilibrium PRESSURE: 401.281 psia — **365.1 psia**
Press RETURN to Continue . . .

Equilibrium Hydrate: STRUCTURE II
Composition of Phases at Equilibrium (moles)

	FEED	WATER	VAPOR	HYDRATE
Methane	0.7329	0.0005	0.7329	0.5779
Ethane	0.0670	0.0001	0.0670	0.0299
Propane	0.0390	0.0000	0.0390	0.3074
i-Butane	0.0036	0.0000	0.0036	0.0408
n-Butane	0.0055	0.0000	0.0055	0.0063
Nitrogen	0.1500	0.0001	0.1500	0.0378
n-Pentane	0.0020	0.0000	0.0020	0.0000
n-Hexane	0.0000	0.0000	0.0000	0.0000

Press RETURN to Continue . . .
Fractional Occupancy of Cages

	SMALL	LARGE
Methane	0.6923	0.0445
Ethane	0.0000	0.0739
Propane	0.0000	0.7602
i-Butane	0.0000	0.1008
n-Butane	0.0000	0.0155
Nitrogen	0.0461	0.0011
n-Pentane	0.0000	0.0000
n-Hexane	0.0000	0.0000

Do you wish to do another calculation at the SAME composition? (Y or N)
N
PLEASE CHOOSE ONE OF THE FOLLOWING OPTIONS
(1) Pressure predication at a given Temperature
(2) Pressure prediction at a given Temperature with an inhibitor
(3) Change Feed Composition
(4) Change Program Units
(5) Return to Main Menu
(6) Quit HYDOFF
5
THE FOLLOWING OPTIONS ARE CURRENTLY AVAILABLE

(1) MAIN Program for Equilibrium Hydrate Predictions
(2) Display Feed Composition
(3) Change Feed Composition
(4) Change Program Units
(5) DISCARD all Data and begin NEW Problem
(6) Exit HYDOFF Program
6
Are you sure you want to exit CSMHYD? (Y or N)
Y

Example B.2: Pressure Prediction With MeOH.[3]

Input Success and Data

HYDRATE PREDICTION PROGRAM: HYDOFF

Press RETURN to continue ...
AVAILABLE UNITS ARE AS FOLLOWS:

	TEMPERATURE	PRESSURE
(1)	Fahrenheit	psia
(2)	Kelvin	kPa

Please select the desired set of Units:
1
The program has been designed to allow the user to input the feed components and composition directly in the program or through an external file, namely, FEED.DAT.

If the user wishes to read the feed components and composition from FEED.DAT, please make sure the information is entered correctly into FEED.DAT (user has to CHANGE the COMPOSITIONS ONLY) and that FEED.DAT is in the same directory as the executable HYDOFF.EXE file.

Are the FEED COMPONENTS and COMPOSITION saved in FEED.DAT (Y or N)?
N
How many COMPONENTS (excluding Water) are present?
7

sI and sII HYDRATE FORMERS

1. Methane	2. Ethane	3. Propane	
4. i-Butane	5. n-Butane	6. Hydrogen Sulfide	
7. Nitrogen	8. Carbon Dioxide		

NON-HYDRATE FORMERS

9. n-Pentane	10. i-Pentane	11. Hexane	12. Heptane
13. Octane	14. Nonane	15. Decane	16. Toluene

Which Components are present? Please list Hydrate formers first.
1 2 3 5 7 8 9

Enter the MOLE FRACTIONS of each Component
Mole Fraction of Methane : 0.7160

Mole Fraction of Ethane: 0.0473

Mole Fraction of Propane: 0.0194

Mole Fraction of n-Butane : 0.0079

Mole Fraction of Nitrogen: 0.0596

Mole Fraction of Carbon Dioxide: 0.1419

Mole Fraction of n-Pentane: 0.0079

THE FOLLOWING OPTIONS ARE CURRENTLY AVAILABLE
(1) MAIN Program for Equilibrium Hydrate Predictions
(2) Display Feed Composition
(3) Change Feed Composition
(4) Change Program Units
(5) DISCARD all Data and begin NEW Problem
(6) Exit HYDOFF Program
1
PLEASE CHOOSE ONE OF THE FOLLOWING OPTIONS
(1) Pressure prediction at a given Temperature
(2) Pressure prediction at a given Temperature with an inhibitor
(3) Change Feed Composition
(4) Change Program Units
(5) Return to Main Menu
(6) Quit HYDOFF
2
Current Salt is: NaCl
Please select the inhibitor(s) to be used.
(1) Methanol
(2) Salt
(3) Methanol/Salt mixture
(4) Specify new pure salt or salt mixture
(5) Return to Previous Menu
1
Enter the WEIGHT PERCENT of Methanol (up to 50wt%)
10
Enter the required Temperature (in F)
47.03

Program Output

FOUR-PHASE (Lw-H-V-Lhc) EQUILIBRIUM CONDITION WITH INHIBITOR(S)
Inhibitor: 10.000 wt% Methanol
Temperature: 47.030 F — **Experimental pressure**
Equilibrium PRESSURE: 747.560 psia — **800.6 psia**
Press RETURN to Continue . . .
Equilibrium Hydrate: STRUCTURE II

Composition of Phases at Equilibrium (moles)

	FEED	WATER	VAPOR	LIQUID	HYDRATE
Methane	0.7160	0.0014	0.7160	0.7160	0.6017
Ethane	0.0473	0.0012	0.0473	0.0473	0.0404
Propane	0.0194	0.0000	0.0194	0.0194	0.2631
n-Butane	0.0079	0.0000	0.0079	0.0079	0.0134
Nitrogen	0.0596	0.0000	0.0596	0.0596	0.0166
Carbon Dioxide	0.1419	0.0058	0.1419	0.1419	0.0648
n-Pentane	0.0079	0.0000	0.0079	0.0079	0.0000

Author's Note: Equal vapor and liquid compositions indicate a single-fluid phase not discriminated between vapor and liquid.
Press RETURN to Continue . . .
Fractional Occupancy of Cages

	SMALL	LARGE
Methane	0.7594	0.1019
Ethane	0.0000	0.1088
Propane	0.0000	0.7086
n-Butane	0.0000	0.0362
Nitrogen	0.0218	0.0010
Carbon Dioxide	0.0680	0.0386
n-Pentane	0.0000	0.0000

Do you wish to do another calculation at the SAME composition? (Y or N)
N

Input Success and Data

PLEASE CHOOSE ONE OF THE FOLLOWING OPTIONS
(1) Pressure prediction at a given Temperature
(2) Pressure prediction at a given Temperature with an inhibitor
(3) Change Feed Composition
(4) Change Program Units
(5) Return to Main Menu
(6) Quit HYDOFF
2
Current Salt is: NaCl
Please select the inhibitor(s) to be used.
(1) Methanol
(2) Salt
(3) Methanol/Salt mixture
(4) Specify new pure salt or salt mixture
(5) Return to Previous Menu
1
Enter the WEIGHT PERCENT of Methanol (up to 50wt%)
20
Enter the required Temperature (in F)
33.71

Program Output

FOUR-PHASE (Lw-H-V-Lhc) EQUILIBRIUM CONDITION WITH INHIBITOR(S)
Inhibitor: 20.000 wt% Methanol
Temperature: 33.710 F — **Experimental pressure**
Equilibrium PRESSURE: 551.081 psia — **691.8 psia**
Press RETURN to Continue . . .
Equilibrium Hydrate: STRUCTURE II

Composition of Phases at Equilibrium (moles)

	FEED	WATER	VAPOR	LIQUID	HYDRATE
Methane	0.7160	0.0015	0.7160	0.7160	0.5920
Ethane	0.0473	0.0018	0.0473	0.0473	0.0366
Propane	0.0194	0.0000	0.0194	0.0194	0.2782
n-Butane	0.0079	0.0000	0.0079	0.0079	0.0140
Nitrogen	0.0596	0.0000	0.0596	0.0596	0.0149
Carbon Dioxide	0.1419	0.0060	0.1419	0.1419	0.0643
n-Pentane	0.0079	0.0000	0.0079	0.0079	0.0000

Author's Note: Equal vapor and liquid compositions indicate a single-fluid phase not discriminated between vapor and liquid.

Press RETURN to Continue . . .

Fractional Occupancy of Cages

	SMALL	LARGE
Methane	0.7588	0.0777
Ethane	0.0000	0.0987
Propane	0.0000	0.7498
n-Butane	0.0000	0.0377
Nitrogen	0.0197	0.0007
Carbon Dioxide	0.0709	0.0314
n-Pentane	0.0000	0.0000

Do you wish to do another calculation at the SAME composition? (Y or N)

N

PLEASE CHOOSE ONE OF THE FOLLOWING OPTIONS
(1) Pressure predication at a given Temperature
(2) Pressure prediction at a given Temperature with an inhibitor
(3) Change Feed Composition
(4) Change Program Units
(5) Return to Main Menu
(6) Quit HYDOFF

5

THE FOLLOWING OPTIONS ARE CURRENTLY AVAILABLE
(1) MAIN Program for Equilibrium Hydrate Predictions
(2) Display Feed Composition
(3) Change Feed Composition
(4) Change Program Units
(5) DISCARD all Data and begin NEW Problem
(6) Exit HYDOFF Program

6

Are you sure you want to exit CSMHYD? (Y or N)

Y

Example B.3: Temperature and Pressure Predictions With Salt(s)[4]

Input Success and Data

HYDRATE PREDICTION PROGRAM HYDOFF

Press RETURN to continue . . .

AVAILABLE UNITS ARE AS FOLLOWS:

	TEMPERATURE	PRESSURE
(1)	Fahrenheit	psia
(2)	Kelvin	kPa

Please select the desired set of Units:

1

The program has been designed to allow the user to input the feed components and composition directly in the program or through an external file, namely, FEED.DAT.

If the user wishes to read the feed components and composition from FEED.DAT, please make sure the information is entered correctly into FEED.DAT (user has to CHANGE the COMPOSITIONS ONLY) and FEED.DAT is in the same directory as the executable HYDOFF.EXE file.

Are the FEED COMPONENTS and COMPOSITION saved in FEED.DAT (Y or N)?

N

How many COMPONENTS (excluding Water) are present?

2

sI and sII HYDRATE FORMERS

1. Methane	2. Ethane	3. Propane	
4. i-Butane	5. n-Butane	6. Hydrogen Sulfide	
7. Nitrogen	8. Carbon Dioxide		

NON-HYDRATE FORMERS

9. n-Pentane	10. i-Pentane	11. Hexane	12. Heptane
13. Octane	14. Nonane	15. Decane	16. Toluene

Which Components are present? Please list Hydrate formers first.

1 8

Enter the MOLE FRACTIONS of each Component:

Mole Fraction of Methane: 0.8470

Mole Fraction of Carbon Dioxide: 0.1530

THE FOLLOWING OPTIONS ARE CURRENTLY AVAILABLE
(1) MAIN Program for Equilibrium Hydrate Predictions
(2) Display Feed Composition
(3) Change Feed Composition
(4) Change Program Units
(5) DISCARD all Data and begin NEW Problem
(6) Exit HYDOFF Program

1

PLEASE CHOOSE ONE OF THE FOLLOWING OPTIONS
(1) Pressure predication at a given Temperature
(2) Pressure prediction at a given Temperature with an inhibitor
(3) Change Feed Composition
(4) Change Program Units
(5) Return to Main Menu
(6) Quit HYDOFF

1

Enter the required Temperature (in F)

40.01

Program Output

THREE-PHASE (Lw-H-V) EQUILIBRIUM CONDITION

Temperature: 40.010 F — **Experimental pressure**

Equilibrium PRESSURE: 496.699 psia — **494.6 psia**

Press RETURN to Continue . . .

Equilibrium Hydrate: STRUCTURE I

Composition of Phases at Equilibrium (moles)

	FEED	WATER	VAPOR	HYDRATE
Methane	0.8470	0.0011	0.8470	0.7222
Carbon Dioxide	0.1530	0.0058	0.1530	0.2778

Press RETURN to Continue . . .

Fractional Occupancy of Cages

	SMALL	LARGE
Methane	0.7737	0.6610
Carbon Dioxide	0.1034	0.3191

Do you wish to do another calculation at the SAME composition? (Y or N)

N

Input Success and Data

PLEASE CHOOSE ONE OF THE FOLLOWING OPTIONS
(1) Pressure predication at a given Temperature
(2) Pressure prediction at a given Temperature with an inhibitor
(3) Change Feed Composition
(4) Change Program Units
(5) Return to Main Menu
(6) Quit HYDOFF

3

Enter the MOLE FRACTIONS of each Component:

Mole Fraction of Methane: 0.823

Mole Fraction of Carbon Dioxide: 0.177

PLEASE CHOOSE ONE OF THE FOLLOWING OPTIONS
(1) Pressure predication at a given Temperature
(2) Pressure prediction at a given Temperature with an inhibitor
(3) Change Feed Composition
(4) Change Program Units
(5) Return to Main Menu
(6) Quit HYDOFF

2

Current Salt is: NaCl

Please select the inhibitor(s) to be used.
(1) Methanol
(2) Salt
(3) Methanol/Salt mixture
(4) Specify new pure salt or salt mixture
(5) Return to Previous Menu

2

Enter the WEIGHT PERCENT of the Salt or Salt Mixture in the free water

5.02
Enter the required Temperature (in F)
47.93

Program Output

THREE-PHASE (Lw-H-V) EQUILIBRIUM CONDITION WITH INHIBITOR(S)
Inhibitor: 5.020 wt% NaCl
Temperature: 47.930 F **Experimental pressure**
Equilibrium PRESSURE: 979.996 psia **1012.4 psia**
Press RETURN to Continue . . .
Equilibrium Hydrate: STRUCTURE I
Composition of Phases at Equilibrium (moles)

	FEED	WATER	VAPOR	HYDRATE
Methane	0.8230	0.0017	0.8230	0.7150
Carbon Dioxide	0.1770	0.0089	0.1770	0.2850

Press RETURN to Continue . . .
Fractional Occupancy of Cages

	SMALL	LARGE
Methane	0.8028	0.6566
Carbon Dioxide	0.1136	0.3305

Do you wish to do another calculation at the SAME composition? (Y or N)
N
PLEASE CHOOSE ONE OF THE FOLLOWING OPTIONS
(1) Pressure predication at a given Temperature
(2) Pressure prediction at a given Temperature with an inhibitor
(3) Change Feed Composition
(4) Change Program Units
(5) Return to Main Menu
(6) Quit HYDOFF
5
THE FOLLOWING OPTIONS ARE CURRENTLY AVAILABLE
(1) MAIN Program for Equilibrium Hydrate Predictions
(2) Display Feed Composition
(3) Change Feed Composition
(4) Change Program Units
(5) DISCARD all Data and begin NEW Problem
(6) Exit HYDOFF Program
6
Are you sure you want to exit CSMHYD? (Y or N)
Y

B.4 XPAND

B.4.1 Program Overview. This program is used to calculate Joule-Thomson cooling of a gas with expansion across a restriction, such as a control valve. Please note that this program can calculate only gas expansions that contain methane, ethane, propane, *n*-butane, *i*-butane, and *i*-pentane. The program will not calculate accurately expansions for gases containing nitrogen, carbon dioxide, or hydrogen sulfide.

B.4.2 Running the Program. The file is located in the floppy that has been attached to this book. To install XPAND, do the following.

1. Insert the disk into the drive.
2. Copy the file XPAND.EXE from the disk to the hard drive.
3. Obtain/copy the file DOSXMSF.EXE to the same hard drive directory.

After copying, to access the program on your computer, you must be in MS-DOS or a Windows MS-DOS prompt. To run XPAND, do the following:

1. Locate the directory that contains XPAND.EXE and DOSXMSF.EXE.
2. Type "XPAND."

The program will run with the initial display "Enter the number of components." Execute the program through the following steps:

1. Enter the number of components in the expanding gas. The value entered must be between 1 and 6.
2. A menu will be displayed listing six different gas components. Select the components that are present in the natural gas by entering the number corresponding to each component and pressing Enter or Return. Continue to do this until all the components in the gas are entered.
3. A screen appears that requests input of the mole fraction of each component specified in the previous screen. After entering each value, press Enter or Return.

Note: The composition of the gas has to be entered on a mole fraction basis and not on a mol% basis.

4. A prompt appears requesting you to enter the following.
   - The upstream pressure before the gas expansion (psia).
   - The upstream temperature before the gas expansion (°R).
   - The downstream pressure (psia).

Press Enter or Return after each entry.

5. A prompt appears that requests input of a first guess of the downstream temperature, *T*, in R°. This guess is the decreased temperature after expansion.

Once *T* is entered, a table appears listing the initial conditions and the *H* across the expansion. For Joule-Thomson cooling, at the correct *T* the *H* across the expansion should be negligible (zero). Consequently, guesses for *T* should be entered until the *H* is within 0.500 Btu/lbm mol. Once this is done, record the XPAND initial and final conditions before pressing Enter to leave the program.

B.4.3 Output and Examples From the Program. This method may be used to get the final temperature on expansion of a gas from an upstream temperature and pressure to a downstream pressure. However, because the expansion curves are not linear in pressure and temperature, repeat this process with the same upstream temperature pressure, but with several intermediate downstream pressures. Plot the $\Delta H = 0$ expansion pressure/temperature line to determine an intersection with the hydrate-formation line, obtained with HYDOFF.

Example B.4: Step-by-Step Calculation of the Gas Expansion Found in Example 2.12, Chap. 2. These steps were used to calculate the final temperature of a gas expanded from 1,500 psia, 100°F, to 300 psia.

Gas Composition:	Component	Mol%
	Methane	92.70
	Ethane	5.30
	Propane	1.40
	i-butane	1.40
	n-butane	0.34
	i-pentane	0.14

Enter the number of Components:
6
Which components are present?
1 = CH4, 2 = C2H6, 3 = C3H8
4 = i-C4H10, 5 = n-C4H10, 6 = i-C5H12
Component 1:
1
Component 2:
2
Component 3:
3
Component 4:
4
Component 5:
5
Component 6:
6
Enter the mole fraction of each component.
Methane:
0.927
Ethane:
0.053
Propane:
0.014
i-Butane:
0.014
n-Butane:
0.0034
i-Pentane:
0.0014
Enter P1 (psia):
1500
Enter T1 (R):
559.7
Enter P2 (psia):
300

Enter your guess for T2 (R) **1st Guess**
(Enter "0" to exit the program).
520

8.336287E-01
9.461145E-01
Pl = 1500.000 psia Tl = 559.700 R
P2 = 300.000 psia T2 = 520.000 R

1st delta H = 891.234 BTU/lbmol
Ideal gas delta H = -376.414 BTU/lbmol
2nd delta H = 201.219 BTU/lbmol
Total delta H = 313.602 BTU/lbmol

1st delta S = .179 BTU/lbmol-R
Ideal gas delta S = 2.501 BTU/lbmol-R
2nd delta S = .059 BTU/lbmol-R
Total delta S = 2.620 BTU/lbmol-R

If the above values are unsatisfactory, enter
another guess for outlet temperature in
degrees Rankine.

Author's Note: Because we are looking for a ΔH value of zero, we will continue guessing in this example until we approximate such a value.

Enter your guess for T2(R) **2nd Guess**
(Enter "0" to exit the program).
500

8.336287E-01
9.377816E-01
Pl = 1500.000 psia Tl = 559.700 R
P2 = 300.000 psia T2 = 500.000 R

1st delta H = 891.234 BTU/lbmol
Ideal gas delta H = -562.102 BTU/lbmol
2nd delta H = 216.348 BTU/lbmol
Total delta H = 112.784 BTU/lbmol

1st delta S = .179 BTU/lbm mol-R
Ideal gas delta S = 2.136 BTU/lbmol-R
2nd delta S = .070 BTU/lbmol-R
Total delta S = 2.245 BTU/lbmol-R

If the above values are unsatisfactory, enter
another guess for outlet temperature in
degrees Rankine.

Author's Note: After several subsequent guesses, the following guess is obtained to make the ΔH value nil ($\Delta H \approx 0$).

Enter your guess for T2(R) **3rd Guess**
(Enter "0" to exit the program).
488.7

8.336287E-01
9.324399E-01
Pl = 1500.000 psia Tl = 559.700 R
P2 = 300.000 psia T2 = 488.700 R

1st delta H = 891.234 BTU/lbmol
Ideal gas delta H = -665.909 BTU/lbmol
2nd delta H = 225.689 BTU/lbmol
Total delta H = -.364 BTU/lbmol

1st delta S = .179 BTU/lbmol-R
Ideal gas delta S = 1.926 BTU/lbmol-R
2nd delta S = .078 BTU/lbmol-R
Total delta S = 2.027 BTU/lbmol-R

If the above values are unsatisfactory, enter
another guess for outlet temperature in
degrees Rankine.
Enter your guess for T2(R)
(Enter "0" to exit the program).
0

The third guess of T = 488.7°R resulted in an XPAND calculation of $\Delta H = -0.364$ Btu/lbm mol for the six-component gas mixture. This value of total ΔH is sufficiently close to zero, indicating an isenthalpic expansion process.

This result indicates that a pressure drop from 1,500 psia, 100°F, to 300 psia causes a gas temperature reduction to 29°F (488.7°R). Several such calculations at intermediate downstream pressures should be done, because the expansion p/T line is nonlinear. The intersection point of the p/T expansion line (obtained from several XPAND calculations) with the hydrate-formation line (obtained from HYDOFF) will differ from the intersection point obtained with use of only a straight line drawn between the two endpoints for the p/T expansion (1,500 psia, 100°F, and 300 psia, 29°F) and the hydrate-formation line.

Nomenclature

I = ice
H = hydrate
ΔH = isenthalpic gas expansion, BTU/lbm mol
L_{HC} = liquid hydrocarbon
L_w = liquid water
p = pressure, m/Lt^2, psia
ΔS = isentropic gas expansion, BTU/lbm mol
T = temperature, T, °F or °R
V = vapor

References

1. Sloan, E.D. Jr.: *Clathrate Hydrates of Natural Gases*, second edition, Marcel Drekker Inc., New York City (1998).
2. Kobayashi, R. *et al.*: "Gas Hydrate Formation With Brine and Ethanol Solutions," *Proc.*, 13th Annual Convention Natl. Gas Assn. of America (1951) 27.
3. Ng, H.-J. and Robinson, D.B.: "Equilibrium Phase Composition and Hydrating Conditions in Systems Containing Methanol, Light Hydrocarbons, Carbon Dioxide, and Hydrogen Sulfide," GPA Research Report RR-66, Gas Processors Assn./Canadian Gas Processors Assn., Tulsa, Oklahoma (April 1983).
4. Dholabhai, P.D. and Bishnoi, P.R.: "Hydrate Equilibrium Conditions in Aqueous Electrolyte Solutions: Mixtures of Methane and Carbon Dioxide," *J. Chem. Eng. Data* (1994) **39**, 191.

SI Metric Conversion Factors

Btu	× 1.055 056	E + 00	= kJ
in.	× 2.54*	E + 00	= cm
°F	(°F − 32)/1.8		= °C
lbm	× 4.535 924	E − 01	= kg
psi	× 6.894 757	E + 00	= kPa
°R	5/9		= K

*Conversion factor is exact.

Appendix C
Additional Case Studies of Hydrate Blockage and Remediation

Case Study C.1[1]

Placid experienced a hydrate-plugging problem in an export pipeline. The prospect was located at Green Canyon Block 29 in the Gulf of Mexico in 1,527 ft of water. A flexible line was installed between the floating production platform to the top of a rigid riser, located 200 ft below the water line. The flexible pipe had a 12-in. inner diameter (ID), a 16-in. outer diameter (OD), and a working pressure rating of 2,160 psi. The export line carried gas and condensate a distance of 52 miles. Flowing conditions before the blockage were 12 MMscf/D of gas and 5,500 BOPD condensate. The oil gravity was 49°API, gas gravity was 0.68, and pipeline inlet conditions were 70°F and 1,050 psi.

During the first few weeks of production, the wells did not produce significant quantities of water. To save operating costs, the gas dehydrators were shut down. When additional wells were brought on stream, some residual water-based completion fluid was being produced. When the wet gas and condensate entered the cold export line (65°F), water condensed and accumulated at the bottom of the catenary loop in the flexible line at 200 ft below the surface. Because the line was not being pigged, water accumulated in this low spot. High-pressure gas combined with cold water in the flexible line to form a complete hydrate blockage in a 14-hour period, causing line pressure to increase to 1,800 psi before production was stopped.

The blockage was located by venting the gas above the plug and filling the void with liquid. The volume of liquid and pressure was recorded. The volume of fluid required to fill the line corresponded to approximately 200 ft of pipeline, suggesting that the blockage was located near the surface. The blockage length was calculated to be 8 to 10 ft long. The export line was depressurized on both sides, and the gas dissociated from the hydrates was vented. The line was pigged successfully with the product gas and condensate the next day. This incident resulted in 3 days of production downtime at an operating cost of U.S. $40,000.

To prevent hydrate formation, three changes were made to the pipeline operations.

1. Methanol was injected.
2. Gas was dehydrated continuously.
3. The line was cleaned periodically with foam pigs.

Case Study C.2[1]

Chevron had a 4-in.-OD, 2,200-ft-long gas flowline plugged with hydrates during the winter. This flowline is in the Whitney Canyon field located in the Carter Creek area of Wyoming. The flowing conditions were 120°F and 360 psig at the wellhead. The ground surface temperature was −20°F, which was well below the hydrate-formation temperature at 360 psig.

The flowline was wrapped with heating tape and insulation to keep the line warm enough to prevent freezing or hydrate formation. Before this blockage occurred, no hydrate inhibitors were used. A corrosion inhibitor was used to prevent corrosion. The line was not equipped for pigging. The line ID was 3.826 in. with a working-pressure limit of 1,800 psi. The flowline material was carbon steel. The heat input was lowered to conserve electrical-energy consumption. However, because there was no mechanism to monitor the fluid temperature throughout the line to ensure that hydrates would not form as the heat input was reduced, a blockage occurred.

A combination of depressurization, chemical, and thermal techniques was used to remove the plug. First, the pressure on both sides of the plug was equalized so that the plug would not move like a projectile. Then, the pressure on both sides of the plug was reduced and methanol was injected upstream of the hydrate plug. The line was heated with the heating tape. This was effective in dissociating the hydrate plug. Production was shut down for 1 day for this remedial operation. Several lessons were learned from this experience. Future operations considered the use of hydrate inhibitors in the winter months. Currently, Chevron is installing pumps to inject a kinetic inhibitor or alternative cost-effective chemicals.

Case Study C.3[1]

In Chevron's platform operations in the Gulf of Mexico, hydrates typically form in the gas-lift distribution valves on the platform. The gas generally is not dehydrated. In the winter, as the gas is throttled through the distribution valve, the Joule-Thomson cooling across the valve drop causes hydrate formation (see Chap. 2, Sec. 2.8). The gas pressure is approximately 1,100 psi. The problem usually is not severe. Because surface access is normally available to the blocked location, methanol can be injected to clear the blockage in the line. To prevent this problem, methanol usually is injected. One solution recently being tested is varying the gas flow rate to keep the valves and gas-distribution lines warm enough to keep them above the hydrate-formation temperature.

Case Study C.4[1]

Chevron reported a hydrate problem in their Carter Knox field in south central Oklahoma. Hydrates formed in an uninsulated 4-in. Schedule 80 (4-in.-ID, 4½-in.-OD) sales gas line. Flowing wellhead conditions were 105°F and 5,750 psi. After choking the well stream to 620 psi at the production unit, the temperature dropped to approximately 62°F at the pipeline inlet. The production unit was designed to remove liquids from the well stream, but the gas was saturated with water vapor and there was always some liquid carryover into the vapor phase. In the winter when the ambient temperature was 45 to 50°F, the gas cooled rapidly owing to the cold environment. Before hydrates formed, no methanol or other chemical was injected at the wellhead or at the processing unit. The well was flowing 200 B/D of oil (57°API) and 7.5 MMscf/D of gas. Water production rate was 10 B/D.

Two flowmeters were installed approximately 120 ft downstream from the production unit on the sales line. One meter is 4-in.-ID with a 2.25-in. orifice plate and the other meter is 3-in.-ID with a 2.125-in.-ID orifice plate. Additional pressure drop occurred under flowing conditions at the second meter, causing hydrates to form there. In fact, the hydrate accumulation near the meter caused an erroneous flow reading that deviated from the first meter. This was an early indicator of the hydrate formation, and it was detected before a complete blockage occurred. It took several hours for the hydrates to form.

To remove the hydrate plug, the line was depressurized and a pump injected methanol into the line. The production unit was preheated to 190°F before startup. It took 4 hours to remove the hydrate accumulation completely. Furthermore, production was shut down for about 8 to 10 hours. On the basis of this experience, methanol injection was begun at the rate of 10 gal/D whenever ambient temperature dropped below 50°F. The operator considered changing the 3-in.-ID flowmeter to a 4-in.-ID flowmeter to eliminate the severe sales-line restriction.

Case Study C.5[1]

Chevron reported several incidents of hydrate blockages in onshore gas-gathering lines in Canada. In one incident, a complete blockage formed in a 6-in., 15-mile-long pipeline. The pipe was rated to a workingpressure limit of 1,000 psi. The line was insulated with a polymer coating, which is sufficient to keep the gas above the hydrate-formation temperature under flowing conditions. The condensate rate was approximately 20 bbl/MMscf. Although there was no free water, the gas was saturated with water vapor at the pipeline inlet pressure and temperature. The condensed water contributed to forming the hydrate plug. Ambient temperature was approximately 37 to 41°F. The blockage occurred during an extended shut-in period over a 300-ft section underneath a road crossing. Previously, hot taps had located a blockage in the same location. While hot tapping was an option, it was considered too risky in this case. Furthermore, hydrates typically did not form in these 6-in. lines if depressurized within the first 24 hours. To remove the blockage, two methods were used simultaneously. First, the line was depressurized on both sides of the plug. Then, a welding rig applied electric current directly to the 300-ft section of the steel pipe. The line was heated to 68 to 77°F with the welding rig. This approach was effective in melting the hydrate plug. The remedial operation took 2 days to complete.

Case Study C.6[1]

Lasmo experienced a wax and hydrate combination in its Staffa field in the U.K. sector of the North Sea in 1993. A single, uninsulated, 8-in. flowline was installed between two satellite wells and a minimum processing platform facility (Ninian Southern platform) located 6.3 miles away. Furthermore, there was no capability for round-trip pigging because a single line was used. The seabed terrain near the tree was uneven and the flowline passed over another flowline about 1.2 to 1.9 miles from the tree.

Production conditions were 6,000 BOPD with a gas/oil ratio (GOR) of 1,600 scf/bbl and a 0.5 to 1% water cut. The produced fluid consisted of a high-GOR, high-gravity crude with some water. Fluids were produced from the reservoir by a pressure-decline mechanism. The average cloud-point temperature of the crude oil was 79°F. The wax content in the crude oil was 5%. The flowing wellhead pressure was 942 to 1,595 psia at temperatures of 122 to 194°F. Because of the very high heat losses to the sea through the uninsulated line, the unseparated multiphase stream cooled to the seabed temperature within 1.2 to 1.9 miles of the tree. The fluid arrived at the platform at a temperature of 44°F, which was well below the wax cloud-point temperature.

Without thorough documentation, it is believed that hydrate formation (owing to erratic methanol injection) might have served as a nucleation point to cause wax precipitation in this line. In any case, wax deposited in the flowline within a period of several days after production was started. Even though certain paraffin inhibitors were used, they were not completely effective. Periodically, the flowline was soaked with chemical solvents without much success. Sometimes, pressure was applied to force the plug, but this exacerbated the problem by accumulating the paraffin into a ball. Thermochemical, heat-generating chemicals were considered but were rejected because they were considered relatively new technology.

As a contingency plan, Lasmo developed an inductive heating coil to be deployed by use of a remotely operated vehicle to heat the flowline and melt the wax inside the line. Although this technique was developed, it was never implemented in the field. Two problems with this technique were that a significant amount of power and time were required to heat the flowline and its contents and that, even after melting the wax and flowing it, it could cool and redeposit before arriving at the platform.

Approximately 1.2 miles of the pipeline filled with a wax blockage was cut out and replaced. Even after replacing the blocked section of the line, the line became plugged with wax a second time. Injection of chemical inhibitors or methanol or solvent soaking did not work.

In 1995, Lasmo abandoned the field because of multiple problems with hydrates and wax. Repeated attempts to clear the blockages with such chemicals as methanol had failed, and the company decided that it was not economical to replace another section of pipe (as was done in 1993) considering the amount of reserves remaining.

Case Study C.7[1]

Texaco experienced a hydrate plug in a 12.75-in. gas-export line at a platform located in Garden Banks 189. The water depth was 725 ft. The line connects to a larger gas-transportation line located on the seafloor. In this case, the gas was not dehydrated sufficiently before the gas was pumped into the export line. As a result, the water vapor condensed and settled out in a U bend at the bottom of the riser. The condensed water that collected at the low spot formed hydrates. In this case, hydrates very rapidly formed a near-complete blockage before it was detected. The line injection pressure rose very quickly.

To remove the hydrate plug, the gas was vented from the platform end and methanol was lubricated down the riser. The line had a check valve downstream of the riser to prevent gas from backflowing to the platform. After injecting some methanol, the hydrates completely melted and the line was cleared. A total of 20 to 30 55-gal drums of methanol was used for the entire operation. Production of 8,000 B/D of liquid and 70 MMscf/D of gas from the platform was shut down for 2 to 3 days during this remedial operation.

Case Study C.8[1]

Texaco reported a hydrate restriction in another gas-export line from a platform at Green Canyon Block 6 in 600 ft of water. Hydrates slowly accumulated in a 10.75-in. line over a period of several days. While production was not shut down, two actions were taken to remove the restriction: (1) the gas dehydrator was turned on to remove water vapor from the gas stream and (2) methanol was injected into the gas-export line.

Case Study C.9[1]

Texaco reported a gas-hydrate blockage in an instrument-isolation valve block in their Strathespay field in the North Sea. However, there were no reports of hydrate blockage in the flowline because the line is insulated adequately. This field was located in 442 ft of water. The valve block had a ¼-in.-ID port leading to a pressure transducer. Because the fluid was static in this section of the line, the produced gas had water vapor that condensed and formed hydrates. The valve block and pressure port were uninsulated and exposed to very cold seawater (39°F). The hydrate blockage resulted in erroneous pressure-transducer readings.

To remove this blockage, the line was purged with methanol. Methanol purging was continued periodically to prevent this problem. This workover operation, however, was undesirable and increased operating cost.

One design flaw with this system was that the transducer line (¼-in.-ID) was situated above the valve block. Even if this line was filled periodically with methanol, the fluid would drain out into the flowline. This allowed wet gas to enter the transducer line and plug it with hydrates. One design option was to change the orientation of the valve block so that the transducer line was connected to the bottom side of the valve block instead of the top side. With this configuration, the line between the flowline and the transducer sensor can be filled with an oil-based gelled fluid mixed with methanol, glycol, or an oil-based fluid. Otherwise, the line may fill with water, causing hydrate formation. In deepwater systems where transducers may be changed as part of a larger system, isolation valves may not be necessary.

Case Study C.10[1]

Elf Norge reported hydrate formation in their North East Frigg subsea flowline. The 16-in. flowline transported gas condensate from a subsea template with six wells, located 11.1 miles from the Frigg platform.

During some period, only one well was flowing at a rate of 35 MMscf/D. At this low gas flow rate, most of the water and condensate settled out and accumulated in the pipeline. After a few days, the gas flow rate was increased by starting three other wells. After the gas flow rate increased to 70 MMscf/D, the pressure and liquid level in the inlet gas/liquid separator became unstable. The wells were shut down. The separator was found filled with hydrates. Samples taken from the separator contained large, solid blocks of hydrates, which took about 1 day to melt.

Analysis of the liquid samples showed that the methanol content was 11 wt%, which was well below the 26 wt% required to avoid hydrate formation. However, Elf reported that the flowline did not plug with hydrates although it experienced subcooling up to 42.8°F. Hydrates were found just downstream of the choke on the platform. Because of Joule-Thomson cooling (see Chap. 2, Sec. 2.8), the gas/water mixture experienced the lowest temperature downstream of the choke. Before restarting production, the separator was depressurized and circulated with steam to remove the hydrates. Approximately 9,000 gal of methanol was injected into the pipeline inlet, the outlet, and upstream of the pipeline outlet choke. An additional 21,000 gal of methanol was injected during the first 2 days of restart when the gas flow rate was gradually re-established.

The liquid outlet valve of the inlet separator was eroded severely during the hydrate-formation period. This might have resulted from a combination of metallic particles, scale, or hydrate crystals flowing at high velocities through the valve. The valve had to be replaced. Another reason hydrates formed downstream of the choke was the lack of an upstream heater. In many subsea completions, a heater is installed upstream of the separator and choke to prevent hydrates or wax formation and to improve separation efficiency.

Case Study C.11[1]

Marathon provided the following information on gas-hydrate formation observed in a gas-export pipeline from their Ewing Bank 873 (EB 873) platform in the Gulf of Mexico.

Hydrate formation occurred in the gas-export line from the EB 873 platform. The line left the platform and contained a 900-ft-deep loop before joining a subsea "T" connection. The line was 8-in. nominal size. The water depth ranged from 775 ft at the EB 873 platform to a maximum of 950 ft, then 470 ft at the subsea connection. Seafloor temperature was estimated to be 55°F. Hydrate formation was inferred from pressure buildup in the line and from the fact that methanol could be used successfully to remediate. Methanol was pumped continuously for inhibition at approximately 140 gal/D for 32 MMscf/D. The pressure drop in the line was a function of flow rate. It normally was in the range of 50 to 100 psi, depending on flow rate, and could be modeled accurately. If it increased much beyond the normal level (e.g. an additional 30 psi), then a slug of methanol was pumped into the line periodically. The hydrate restriction appeared to be between the EB 873 platform and the low point. Pigging was not attempted for a variety of reasons, but primarily because of high risk for minimal benefit. Methanol was a cheap and a low-risk solution. The routine technique of depressurizing the line was not used at EB 873 because shutting in production would be required.

Case Study C.12[1]

Phillips reported gas-hydrate-plugging problems in its Cod pipeline in the North Sea. The 47-mile-long pipeline of 16-in. (ID = 15.125 in.) carbon steel was designed to transport gas and gas condensate from the Cod platform to the Ekofisk center. The liquid was a light hydrocarbon with a specific gravity of 0.66. Cod production was approximately 35 MMscf/D of gas and 1,700 B/D of condensate.

Gas hydrates completely plugged the Cod pipeline several times. In March 1978, hydrates formed and a pig became stuck in the hydrate accumulation. The hydrates were removed by depressurizing the line. The line was backflowed in an unsuccessful attempt to remove the pig. While the pig remained in the line, the restriction did not prevent the gas flow. A slug of 1,700 gal of methanol was pumped to try to dissolve all the hydrates in the line. During the restart, methanol was injected continuously into the pipeline.

On the Cod platform, even though the gas stream was dried adequately, the liquid condensate stream was not. Therefore, the wet condensate stream mixed with the high-pressure gas to form hydrates in the pipeline. Since 1981, the operating pressure has declined so that the pipeline is now operating outside the hydrate-formation conditions.

Case Study C.13[1]

Texaco performed field tests in several of their Wyoming wells to evaluate the use of polyvinylpyrrolidone (PVP), a kinetic inhibitor (see Sec. 2.8). The kinetic inhibitor could be used at very low concentrations, ranging from 0.5 to 1 wt%, instead of 10 to 50 wt% methanol to achieve the required level of hydrate inhibition.

Before the field tests, these Wyoming wells and surface flowlines were experiencing hydrate-plugging problems in the wells at methanol injection rates of 30 gal/D. Flowing wellhead conditions were up to 2,000 psi and 52 to 56 °F. Gas production ranged from 0.8 to 1.4 MMscf/D. Freshwater production rate ranged from 2 to 40 B/D.

Methanol was replaced with a 4% PVP solution. The 4% PVP solution consisted of 4 wt% PVP, 16 wt% water, and 80 wt% methanol. The PVP solution was pumped at a rate of 2 to 21 gal/D, representing an aqueous-phase concentration of less than 0.05 wt%. At these concentrations, the kinetic inhibitor was effective in preventing hydrates. This use of kinetic inhibition represents a cost savings of about 50% compared with the cost of inhibiting with only methanol.

Case Study C.14[1]

Similar to Case Study C.13, Texaco conducted another series of field tests in east Texas to evaluate PVP. In this field, tests were conducted on 4- to 6-in. flowlines that were 1 to 8 miles long. The gas flow ranged from 1 to 24 MMscf/D. Water flow ranged from 0.8 to 40 B/D. Similar to the Wyoming field tests, hydrates formed rapidly when the methanol rate was greatly reduced. Following depressurization, subsequent hydrate plugging was prevented by injecting the kinetic inhibitor at concentrations in the range 0.1 to 0.5 wt% of the aqueous phase.

Texaco has completed extensive testing of kinetic gas-hydrate inhibitors in onshore U.S. fields. Kinetic inhibitors currently are being used in many of their fields to reduce methanol consumption costs. Texaco is continuing to experiment with alternative chemicals for optimizing costs and for application in offshore flowlines.

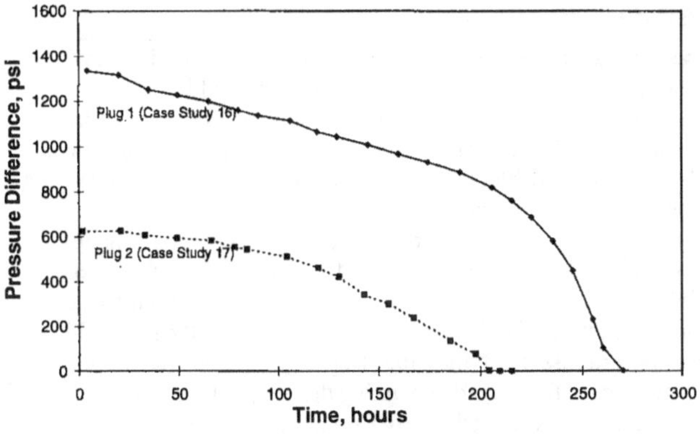

Fig. C.1—Pressure difference across two Tommeliten plugs in Case Studies 16 and 17.[5]

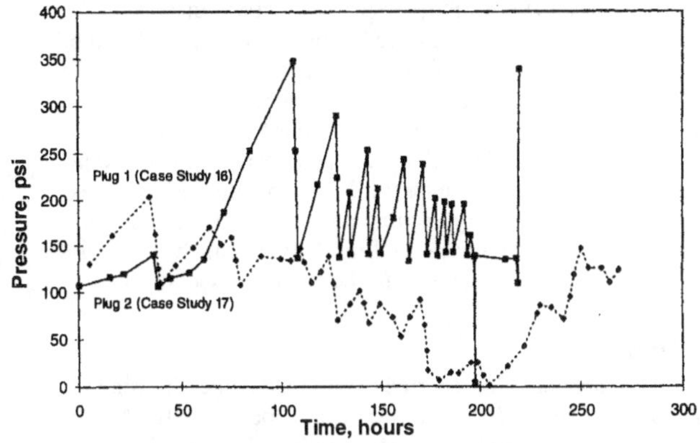

Fig. C.2—Pressure at riser discharge with two Tommeliten plugs in Case Studies 16 and 17.[5]

Combined Case Study C.15

Statoil conducted 19 controlled field experiments of gas-hydrate-blockage formation and dissociation.[2,3,*] The experiments were done in 1994 with a 6-in. test/service subsea line in their Tommeliten Gamma field. The line is connected to the Edda platform, located 7.1 miles away from the subsea manifold. Two 9-in. production lines and one 6-in. test/service line were installed to carry the flow from a subsea production manifold. The manifold gathered the flow from six subsea wells. Condensate content was 16 wt% and water content was 2 wt%.

Nineteen hydrate-formation and -dissociation experiments were conducted with the 6-in. test/service line in three types of experiments as follows.

1. Continuous flow. Statoil lowered the flowing temperature by reducing the flow rate so that the system cooled and entered the hydrate-formation region.
2. Continuous flow without methanol injection. Production rate was reduced and methanol injection was stopped.
3. Restart after shut-in with the following four approaches.

Cooled the pressurized line; restarted without methanol injection.

Cooled the pressurized line; restarted with 5 wt% methanol injection.

Pressurized the line from the template side; restarted at high flow rate.

Pressurized the line from the platform side; restarted at high flow rate.

During these experiments, Statoil measured pressure and temperature at the following places: (1) at the manifold, (2) at the top of the riser upstream of the heater, (3) at the choke, and (4) in the separator. Statoil also used two gamma densitometers to detect the arrival of slugs and hydrate lumps on the platform. A thermocamera was used to detect the temperature profile of the topside lines and to detect ice/hydrate formation.

The following are general conclusions reported by Statoil on those field experiments.

1. Hydrates formed easily and rapidly after fluid conditions entered the hydrate-formation region. In some cases, hydrate chunks flowed to the platform and plugged the topside piping, valves, and bends.
2. Underinhibition of methanol increases rate of hydrate formation and risk of plugging. Field tests were done at 5 wt% methanol. Laboratory tests performed with 10 to 20 wt% methanol also found similar results.[4]
3. Hydrate plugs were porous and permeable. When the plug was subjected to a differential pressure, the gas from the manifold side flowed through the plug. This was indicated by a gradual drop in pressure at the manifold when the gas was being vented from the platform side. See Case Study 3.3 for a plug that was less permeable to a Statoil black oil.
4. Gas flow through the plug caused Joule-Thomson cooling, leading to additional hydrates or ice. If additional hydrates or ice formed in pore spaces within the hydrate plug, the dissociation rate was reduced.
5. Combinations of depressurization and methanol injection were effective to remove all plugs. Methanol could be injected at either manifold end or platform end.
6. Methods to remove hydrates in the topside piping include injecting methanol and/or spraying warm water on the outside surface. However, heating the pipe from the outside can be risky. If the gas released from hydrate dissociation is not vented properly, the trapped gas potentially may overpressure the line.
7. The results and recommendations developed from these field experiments cannot be applied directly to other fields with different conditions and fluid compositions.

*Personal communication with T. Austvik, Statoil Research Center, Trondheim, Norway, 13–15 July 1997.

Case Studies C.16 and C.17

These two case studies summarize two blocking events in Combined Case Study 15.

Case Study C.16. The experiment originated as a depressurized line that was brought into production. Methanol was injected continuously into the line throughout the startup process to prevent hydrate formation. When production reached a rate of 12 MMscf/D, methanol injection was stopped, allowing hydrates to form at the temperature of 60°F. The riser temperature was 16°F below the hydrate-formation region. After several partial blocking events, a complete hydrate plug formed approximately 2.5 miles from the platform (26 hours after startup). The plug location was estimated from evaluating the rate of pressure change on both plug sides.

When blockage occurred, the pipeline was depressurized to dissociate the hydrate plug. In addition, 3,400 gal of methanol was injected into the wellhead to assist in dissociation. Owing to the fact that methanol had to travel 5 miles, the horizontal nature of the pipeline, small pipeline buckling, and liquid present in the pipeline, it is believed that methanol never reached the plug. One-sided depressurization of the pipeline removed the plug after 7 days. Total blockage time was 25 days.

Case Study C.17. The uninhibited line was shut in at full well pressure and cooled to ambient sea temperature. The line was then started and began producing at a rate of 12 MMscf/D without any methanol present. The production line was maintained for 40 hours without any hydrate blockage of the line. Several blocking events occurred topside before a blockage formed somewhere between the template and riser. After observing pressure changes on both sides of the plug, it was determined that the plug was approximately 2.5 miles away from the platform. The hydrate plug was removed through one-sided depressurization. The hydrate plug dissociated slowly, requiring 9 days to be removed.

Fig. C.1 shows the measured pressure difference across the two plugs in Case Studies C.16 and C.17 as a function of time.[5] These curves were generated with large pressure fluctuations that occurred while reducing the pressure removed. The figure highlights the change of permeability of the plug as a function of time.

Fig. C.2 shows the pressure in the riser during the hydrate-removal process.[5] The equilibrium pressure for the hydrate plugs was approximately 200 psi at the ambient temperature. Plug 1 was kept under the equilibrium temperature until it dissociated. Plug 2 was kept temporarily above the equilibrium point to limit the cooling effects caused by Joule-Thomson cooling. It was thought that this practice had little effect on increasing the rate of dissociation.

Case Study C.18

Occidental Oil and Gas Co. reported hydrate blockages forming in a gas and associated condensate transport line located in the North Sea. Hydrate plugs usually formed in subsea interfield pipelines and in the bottom of incoming risers. The export pipeline operated at 4,930 psig with a wellhead temperature of 86°F, which cooled to the ambient sea temperature of 35°F at the pipeline outlet. The cold temperatures placed the pipeline within hydrate-formation conditions for the gas. To combat this, methanol was injected to maintain 25 wt% in the free-water phase.

Hydrates formed when insufficient amounts of methanol were injected into the pipeline. Early symptoms of hydrate formation were increases in differential pressure and reductions in gas production. A late symptom of hydrate plugs was complete blockage of flow. When blockages occurred in the pipeline, two methods were used to remove the plugs. The first method consisted of methanol injection and depressurization of the pipeline from both sides; the usual time needed to remove blockages through this method was 0.5 to 1 day. Depressurization could be avoided by adding large volumes of methanol until dissociation occurred; the usual time needed to carry out this remediation was 4 to 14 days.

Occidental also emphasized the importance of minimizing the differential pressures across the plug to prevent hydrate projectiles. Secondly, the company emphasized that the pressure must be maintained above 87 to 145 psig. If the pressure dropped below these values, the equilibrium temperature moved well below 32°F, causing ice formation. Ice could not be dissociated through depressurization and possibly could require more time to remove than hydrate plugs.

Case Study C.19

Amoco reported hydrate-plug formation in a 70-mile-long export pipeline located in the North Sea. Under normal operating conditions, the gas is dehydrated and then compressed from 350 to 1,300 psig. The concentration of water in the gas phase usually is low enough to prevent free-water formation. However, the line had not been pigged for 3 months and offshore process upsets were thought to have allowed free water into the line during that time. High pressure drops formed in the pipeline, requiring pigging, but the pig became stuck in the line and had to be removed through flow reversal. Hydrate slush appeared with the pig on the offshore platform. After pigging, the pipeline became completely blocked with hydrates approximately 30 miles from the offshore platform. The amount of gas used to displace the pig was used to estimate the plug's location.

Two possible tools for pipeline remediation were methanol injection and depressurization of the pipeline. Methanol could not be used as a remediation method because of the plug's distant location; consequently, two-sided depressurization became the only viable means of dissociating the hydrate plug. Depressurization was carried out over a 2-week period in slow steps to prevent any high pressure buildups owing to multiple plugs. After 8 weeks, the plug was dissociated completely and full production resumed. The line was restarted by slowly sweeping the pipeline with dry gas, building up to high gas rates. The line was consistently pigged, first with undersized pigs and then with full-sized ones. No problems were witnessed during startup.

The hydrate-remediation process lasted 8 weeks and cost U.S. \$500,000 to carry out. Overall, the plug shut down production for 3 months and cost U.S. \$5.5 million because of remediation expenses and loss of sales.

Case Study C.20

Petrobrás reported a hydrate blockage in a subsea manifold, located at approximately 2,000 ft water depth. The manifold initially was loaded with water and was not drained and loaded with ethanol before production startup, the normal practice. Consequently, a hydrate plug formed in the manifold, blocking valves in a production line. However, production was maintained through a test production line.

Two methods were attempted to dissociate the plug. First, ethanol was injected into the manifold to begin dissociation. Some dissociation did occur (indicated by pressure increases), but the hydrate plug was still present after 2 days. Depressurization of the manifold was then used to dissociate the plug. Depressurization was carried out on both sides of the plug, dissociating the plug in 12 hours. Start-up of the pipeline was carried out by filling the manifold with ethanol and then resuming production.

Overall, the hydrate plug was in the manifold for 60 days, but production was maintained throughout that time by a test production line. During depressurization, all production from the wells flowing into the manifold had to be shut down. The total economic loss owing to the hydrate was 31,500 bbl oil and the wages of two engineers (1 week) and two technicians (3 days).

Case Study C.21

Barker and Gomez[6] described an Exxon experience with a hydrate plug in a well located in 1,150 ft of water off the California coast. While drilling, gas flowed into the well from the formation, channeling through the primary cement column at 7,750 ft, and the migrating gas entered the freshwater mud at the subsea wellhead. Because of difficulties with the wellhead-hanger packoff, the gas influx was stopped by perforating the casing, which resulted in severing the drillstring and stripping it up through the blowout preventers (BOP's) until the severed drillstring end was above the gas sand.

A through-drillstring perforating gun was then run to shoot the 7-in. casing just above the gas sand. The gas influx was killed by pumping a 14.2-lbm/gal mud down the drillstring and into the formation at surface pressures up to 3,100 psi. At the conclusion of the kill operation, both the chokeline and the kill line were found plugged. Subsequent operations were hampered by the inability to use either line. After cementing operations that secured the wellbore, the BOP's were recovered. Hydrates and trapped gas were found in the choke and kill lines of the bottom eight riser joints.

Case Study C.22

Barker and Gomez[6] reported a second Exxon drilling instance in 3,100 ft of water in the Gulf of Mexico, with a ocean-bottom temperature of 40°F. Gas flowed into the well and plugged the choke and kill lines. After 4 days of warm-drilling-mud circulation, the lower-middle ram-type BOP's could not be opened or closed completely, possibly because of hydrates in the ram-block recesses.

The drillstring was perforated about 400 ft above the annular gas/liquid contact. After coiled tubing was run inside the drillstring, hot mud was circulated and gas was allowed to migrate into the coiled-tubing/drillstring annulus before being circulated out of the well. Three sets of successively shallower perforations were required to remove the gas in the annulus completely.

After all ram-type BOP's were opened, the drillstring was backed off at 5,000 ft and recovered and a cement plug was set in the casing. The well was secured and the BOP's pulled, resulting in a recovery of hydrates. Testing of BOP's at the surface indicated that the failure was not caused by mechanical failure of the BOP's, which were then free of hydrates.

Case Study C.23

Davalath and Barker[7] described a hydrate problem in 595 ft of water offshore South America. The well was completed with a 7-in. casing and 3.5-in. tubing. Production was gas and condensate at several hundred B/D with a water cut of about 6%. A 15-hour production test was followed by a 25-hour shut-in period to collect reservoir pressure-buildup data.

TABLE C.1—PLUG CHARACTERISTICS IN CASE STUDIES 25–27.

	Test			
	1	2	3	4
Dates, 1997	27–31 January	1–5 February	6–8 February	19–20 February
Block time, hours	85	62	37	143
Plug length, ft	NA	NA	25	90, 300, 30, 70
Maximum dissociation pressure, psi	174	Aborted	390	475
Maximum gas superficial velocity, in./sec	4.75	NA	NA	0.45
Maximum load, lbf	2,216	NA	665	6,050
Leakage, mass/load [(lbm – sec)/lbf]	1.47×10^{-5} to 6.57×10^{-5}	NA	2.8×10^{-6}	3.7×10^{-6}
Maximum plug velocity, ft/sec	NA	NA	270	65
Shear stress, lbf/in.	.074	NA	1.22	1.308

The well was shut in at the surface, which exposed the tubing to high-pressure gas and cold 45°F water, which led to the formation of hydrates. Under these conditions the tubing fluid was about 29°F below the hydrate-formation temperature. Wireline tools were blocked by a bridge inside the tubing string, and further pulling caused separation. Subsequently, the lubricator was found to be full of hydrates.

Attempts were made to melt hydrates by (1) pouring glycol into the top of the tubing, (2) using heated mud and seawater, and (3) increasing the pressure up to 7,000 psi at the surface to break the hydrate plug. These attempts were unsuccessful and the authors noted that the pressure increase caused a more stable hydrate rather than blowing it from the tubing.

A coiled-tubing string was stripped inside the tubing and 175°F glycol was circulated to the hydrate plug at 311 ft. Direct contact with the hot glycol removed the hydrate plug, but more than 13 days were lost because of this incident.

Case Study C.24

Davalath and Barker[7] also reported hydrate formation during well abandonment in the Gulf of Mexico. During normal production, methanol was injected at the subsea tree. After production was stopped, the flowlines and tree piping were filled with seawater and corrosion inhibitor from the surface to the seafloor. During plug-and-abandonment operations, the operator found ice-like solids inside the tubing bore of the tree at the seafloor and in the annulus bore. The solid hydrate plugs were dissolved by circulating heated $CaBr_2$ brine through a coiled-tubing string run inside the tubing.

Case Studies C.25, 26, 27

Three controlled hydrate field tests were completed on Devon Energy Kerr-McGee 900-psia gas-condensate line in the Powder River basin of Converse County, Wyoming, from 27 January to 20 February 1997.[8] The object of the tests was to show that one-sided depressurization can be performed safely in the field. As indicated in the hydrate-plug-remediation portion (Chap. 2) of this book, the standard onshore dissociation procedure is to balance the pressure on both sides and to reduce the balanced pressure to move outside of the hydrate region.

The test line was 4 in., 17,381 ft long from wellhead to separator-receiver unit (SRU-10) and pig receiver and mostly buried to a depth of 5 ft at a ground temperature of 34°F. Elevation varied over 250 ft. Normally in winter, the flowline is treated continually with methanol and pigged daily to prevent hydrate problems.

The pipeline had the following instruments at five sites: (1) the wellhead (Werner-Bolley) with pressure/temperature sensors, 1.5-in. flow orifice, backpressure control valve, and pig launcher; (2) 3,7852 ft downstream with pressure/temperature sensors and blowdown; (3) 5,395 ft downstream with pressure/temperature sensors; (4) 6,624 ft downstream with pressure/temperature sensing, methanol injection, blowdown capability, and dual gamma ray sensors to monitor plug velocity, length, and density (it was difficult to discern the differences between water, plugs, and condensate); and (5) 11,483 ft downstream with pressure/temperature sensing. At the end of the line was SRU-10, which contained a pig receiver and blowdown.

Temperatures were not analyzed because the thermocouple was external to the pipe. However, the pressures at the four downstream sites, the orifice measurement, and the gamma ray measurements proved invaluable in analyzing hydrate formation and dissociation.

The following steps were used to conduct a test.

1. Data collection initiated.
2. Methanol injection stopped at the wellhead.
3. Methanol injection begun at Site 4.
4. Pig launched at Site 1 and received at SRU-10.
5. Blockage formation monitored.
6. Line isolated after blockage formation.
7. Blockage dissociation by blowdown at Sites 2, 4, or SRU-10.

The average steady-state liquid holdup at Site 4 was 3.9%. The liquid in the water/condensate plugs was between 4.5 and 4.9%. Average superficial gas velocity was 6.3 ft/sec in the pipeline, without blockages.

Case Study C.25. Test 1 had two blockages. One relatively impermeable blockage was formed in the cold portion of the line between Sites 4 and 5. The other, more permeable blockage was formed in the warm portion of the line. Both were cleared by blowdown at Site 4. The differential pressure across the blockages ranged between 112 and 174 psi, corresponding to a total load between 1,416 and 9,800 lbf.

Test 2 was aborted because hydrates formed upstream of Site 2 (undesirable from a safety standpoint because Site 2 is above ground with two ball valves and two 45° bends). Hydrates were dissociated by reducing and balancing the pressures on either side.

Case Study C.26. Test 3 resulted in a short (25-ft) blockage with low permeability, which dislodged and passed Site 4 with a speed of 270 ft/sec, before eroding further down the pipeline. The differential pressure ranged from 271 to 475 psi, corresponding to a total load between 3,439.6 and 6,047 lbf, about double that of more porous blockages.

Case Study C.27. Test 4 had a blockage that formed on the downstream side of Site 4 and then was moved upstream of Site 4 by a line depressurization at Site 2. Each time the plug was driven past Site 4, then lodged to form another, less-permeable blockage. These plugs were longer (approximately 90 and 175 ft) than those of Test 3.

Table C.1 summarizes the characteristics of the plugs.

References

1. Davalath, J.: "Methods To Clear Blocked Flowlines," DeepStar IIA CTR A208-1, Mentor Subsea (January 1996).
2. Austvik, T. *et al.*: "Tommeliten Gamma Field Hydrate Experiments," *Proc.*, Intl. BHRA Conference on Multiphase Production, Cannes, France (7–9 June 1995) 539.

3. Austvik, T. *et al.*: "Formation and Removal of Hydrate Plugs, Field Trial at Tommeliten," *Proc.*, Gas Processors Assn. Conference, San Antonio, Texas (1997) 205.
4. Yousif, M.H. *et al.*: "The Effects of Low Concentration Methanol Solutions on Hydrate Formation," *Proc.*, Intl. Conference on Natural Gas Hydrates, Toulouse, France (1996) 291.
5. Berge, L., Gjertsen, L.H., Lysne, D.: "The Importance of Porosity and Permeability of Hydrate Plugs in Pipes," *Proc.*, Intl. Conference on Natural Gas Hydrates, Toulouse, France (1996) 533.
6. Barker, J.W. and Gomez, R.K.: "Formation of Hydrates During Deepwater Drilling Operations," *JPT* (March 1989) 297; *Trans.*, AIME, **287.**
7. Davalath, J. and Barker, J.W.: "Hydrate Inhibitor Design for Deepwater Completions," *SPEDC* (June 1995) 115.
8. Hatton, G.J., Barajas, A.M., and Kuhl, C.A.: Hydrate Plug Decomposition Test Program, Report Deepstar 3204, SWRI Project 04-8217 (October 1997).

SI Metric Conversion Factors

°API	141.5/(131.5 + °API)		= g/cm^3
bbl	× 1.589 873	E − 01	= m^3
°F	(°F − 32)/1.8		= °C
ft	× 3.048*	E − 01	= m
ft^3	× 2.831 685	E − 02	= m^3
gal	× 3.785 412	E − 03	= m^3
in.	× 2.54*	E + 00	= cm
lbf	× 4.448 222	E + 00	= N
lbm	× 4.535 924	E − 01	= kg
mile	× 1.609 344*	E + 00	= km
psi	× 6.894 757	E + 00	= kPa

*Conversion factor is exact.

Appendix D
Rules-of-Thumb Summary

This appendix presents a summary of rules of thumb used in the book along with the section in which they were presented. These rules-of-thumb are based on experience and are intended as guides for further action. For example, guided by a rule of thumb, an engineer might determine that a more accurate calculation is needed for inhibitor-injection amounts or that further consideration of hydrates is unnecessary. Rules of thumb are not intended to be absolute truths, and exceptions can always be found; where possible, the appropriate section provides the accuracy of the rule of thumb.

Rule-of-Thumb 1. (Sec. 2.5) At 39°F, hydrates will form in a natural gas system if free water is available and the pressure is greater than 166 psig.

Rule-of-Thumb 2. (Sec. 2.6) For long pipelines approaching the ocean-bottom temperature of 39°F, **Table D.1** gives the lowest water content of the outlet gas.

Rule-of-Thumb 3. (Sec. 2.6) At 39°F and pressures greater than 1,000 psia, the methanol (MeOH) lost to the vapor phase is 1 lbm MeOH/MMscf for every weight percent MeOH in the free-water phase.

Rule-of-Thumb 4. (Sec. 2.6) At 39°F and pressures greater than 1,000 psia, the maximum amount of monoethylene glycol (MEG) lost to the gas is 0.02 lbm/MMscf.

Rule-of-Thumb 5. (Sec. 2.6) MeOH concentration dissolved in condensate is 0.5 wt%.

Rule-of-Thumb 6. (Sec. 2.6) The mole fraction of MEG in a liquid hydrocarbon at 39°F is 0.03% of the water-phase mole fraction of MEG.

Rule-of-Thumb 7. (Sec. 2.8) Natural gases cool on expansion from pressures less than 6,000 psia; at more than 6,000 psia, the temperature increases on expansion. Virtually all offshore gas processes cool on expansion because only a few reservoirs and no current pipelines or process conditions are at pressures greater than 6,000 psia.

TABLE D.1—GAS WATER CONTENT AT 39°F				
Pipe Pressure, psia	500	1,000	1,500	2,000
Water Content, lbm/MMscf	15.0	9.0	7.0	5.5

Rule-of-Thumb 8. (Sec. 2.8) It is always better to expand a dehydrated gas than a moist gas to prevent hydrate formation.

Rule-of-Thumb 9. (Sec. 2.8) Where drying is not a possibility, it is always better to take a large pressure drop at a process condition where the inlet temperature is high.

Rule-of-Thumb 10. (Sec. 2.9) MEG injection is used when the required MeOH injection rate exceeds 30 gal/hr for onshore gas lines.

Rule-of-Thumb 11. (Sec. 2.9) Use of antiagglomerants requires a substantial oil/condensate phase. The maximum water/oil ratio (volume basis) for the use of an antiagglomerant is 40:60.

Rule-of-Thumb 12. (Sec. 2.9) Polyvinylpyrrolidone may be used to inhibit pipelines with subcooling less than 10°F for flowlines with short gas residence times (less than 20 minutes).

Rule-of-Thumb 13. (Sec. 2.9) VC 713, PVCap, and copolymers of PVCap can be used to inhibit flowlines at subcooling less than 15°F, with water phase residence times up to 30 days.

Rule-of-Thumb 14. (Sec. 2.10) The water and the water-soluble drilling-fluid components determine the conditions of hydrate formation. Concentration of salts, alcohols, glycols, and glycerols in the water determine the hydrate-formation temperature and pressure.

Rule-of-Thumb 15. (Sec. 3.1) Hydrate blockages occur owing to abnormal operating conditions, such as well tests with water, loss of inhibitor injection, dehydrator malfunction, startup, and shut-in. In all recorded instances, pipeline plugs caused solely by hydrates were removed when proper safety precautions were observed.

Rule-of-Thumb 16. (Sec. 3.2) In gas/water systems, hydrates tend to form on the pipe wall. In gas/condensate or gas/oil systems, hydrates frequently form from free water as particles that agglomerate and bridge as larger masses in the bulk streams.

Rule-of-Thumb 17. (Sec. 3.2) Agglomeration of individual hydrate particles causes an open hydrate mass that has a high porosity (typically > 50%) and is permeable to gas flow (permeability/length ratio of 8.7 : 11 × 10^{-15} m). Such an open hydrate mass has the unusual property of transmitting pressure while being a substantial liquid-flow impediment. Hydrate particles anneal to lower permeability over longer times.

Rule-of-Thumb 18. (Sec. 3.3) A lack of hydrate blockages does not indicate a lack of hydrates. Frequently, hydrates form but flow with an oil/condensate (e.g., in an oil with a natural dispersant present) so they can be detected in pigging returns.

Rule-of-Thumb 19. (Sec. 3.3) Attempts to blow the plug out of the line by increasing the pressure differential results in more hydrate formation because higher pressures place the system farther into the hydrate-formation region. When a hydrate blockage is experienced in an offshore system, for safety reasons, the first step is to inject inhibitor into the line from the platform in an attempt to determine the plug distance from the platform. In an onshore system, inhibitor is injected from any access point.

Rule of Thumb 20. (Sec. 3.4) Regardless of the method(s) used, the time required for hydrate dissociation is usually days or weeks. After a deliberate dissociation action is taken, both confidence and patience over a long period of time are required to observe the result.

Rule of Thumb 21. (Sec. 3.4) When dissociating a hydrate plug, it always should be assumed that multiple plugs exist, from both safety and technical standpoints. While one plug might cause the initial blockage, a shut-in causes the entire line to cool rapidly into the hydrate region and low-lying points of water accumulation to convert rapidly to hydrate at the water/gas interfaces.

Rule of Thumb 22. (Sec. 3.4) Because the ends of a hydrate plug cannot be located easily subsea, nonuniform heating is not recommended for subsea hydrate dissociation.

Rule-of-Thumb 23. (Sec. 4.3) MeOH-loss costs can be substantial when the total fraction of either the vapor or the oil/condensate phase is very large relative to the water phase.

Rule-of-Thumb 24. (Sec. 4.3) The 1997 fixed-leg North Sea platform cost was U.S. $77,000/ton.

Rule-of-Thumb 25. (Sec. 4.3) To achieve a desired heat-transfer coefficient of 0.3 Btu/hr-ft-°F, a nonjacketed system costs U.S. $1.5 million/mile. Typical costs of insulation by bundled lines are U.S. $1.5 to $2 million/mile.

SI Metric Conversion Factors

Btu	× 1.055 056	E+00	= kJ
°F	(°F–32)/1.8		= °C
ft	× 3.048*	E–01	= m
gal	× 3.785 412	E–03	= m^3
lbm	× 4.535 924	E–01	= kg
mile	× 1.609 344*	E+00	= km
psi	× 6.894 757	E+00	= kPa
ton	× 9.071 847	E–01	= Mg

Author Index

L

M

N

O

P

R

S

T

U

W

X

Y

Subject Index

D

E

F

G

H

I

J

K

L

M

N

O

P

Q

R

S

T

U

V

W

www.ingramcontent.com/pod-product-compliance
Lightning Source LLC
LaVergne TN
LVHW082356200926
848481LV00012B/3581
9781555630881